HIGH DYNAMIC RANGE VIDEO

HIGH DYNAMIC RANGE VIDEO

Concepts, Technologies, and Applications

Edited by

ALAN CHALMERS
University of Warwick, Coventry, United Kingdom

PATRIZIO CAMPISI
Roma Tre University, Rome, Italy

PETER SHIRLEY
NVIDIA, Salt Lake City, UT, United States

IGOR G. OLAIZOLA
Vicomtech-IK4, San Sebastián, Spain

ELSEVIER

AMSTERDAM • BOSTON • HEIDELBERG • LONDON
NEW YORK • OXFORD • PARIS • SAN DIEGO
SAN FRANCISCO • SINGAPORE • SYDNEY • TOKYO
Academic Press is an imprint of Elsevier

Academic Press is an imprint of Elsevier
125 London Wall, London EC2Y 5AS, United Kingdom
525 B Street, Suite 1800, San Diego, CA 92101-4495, United States
50 Hampshire Street, 5th Floor, Cambridge, MA 02139, United States
The Boulevard, Langford Lane, Kidlington, Oxford OX5 1GB, United Kingdom

Library of Congress Cataloging-in-Publication Data
A catalog record for this book is available from the Library of Congress

British Library Cataloguing-in-Publication Data
A catalogue record for this book is available from the British Library

ISBN: 978-0-12-809477-8

For information on all Academic Press publications
visit our website at https://www.store.elsevier.com/

Publisher: Joe Hayton
Acquisition Editor: Tim Pitts
Editorial Project Manager: Charlotte Kent
Production Project Manager: Susan Li
Designer: Vicky Pearson Esser

Typeset by SPi Global, India

CONTENTS

CONTRIBUTORS

A.O. Akyüz
Middle East Technical University, Ankara, Turkey

A. Artusi
University of Girona, Girona, Spain

F. Banterle
CNR-ISTI, Pisa, Italy

T. Bashford Rogers
University of Warwick, Coventry, United Kingdom

M. Bessa
INESC TEC, Porto, Portugal; Universidade de Trás-os-Montes e Alto Douro, Vila Real, Portugal

M. Čadík
Brno University of Technology, Brno, Czech Republic

A. Chalmers
University of Warwick, Coventry, United Kingdom

Y. Chrysanthou
University of Cyprus, Nicosia, Cyprus

K. Debattista
University of Warwick, Coventry, United Kingdom

J. Gorostegui
Vicomtech-IK4, San Sebastián, Spain

J. Hasić-Telalović
International University of Sarajevo, Sarajevo, Bosnia and Herzegovina

K. Karađuzović-Hadžiabdić
International University of Sarajevo, Sarajevo, Bosnia and Herzegovina

B. Karr
University of Warwick, Coventry, United Kingdom; Rockledge Design Group, Inc., Rockledge, FL, United States

N. Ladas
University of Cyprus, Nicosia, Cyprus

C. Loscos
University of Reims Champagne-Ardennes, Reims, France

R. Mantiuk
West Pomeranian University of Technology, Szczecin, Poland

Á. Martín
Vicomtech-IK4, San Sebastián, Spain

I. Martin
University of Girona, Girona, Spain

M. Melo
INESC TEC, Porto, Portugal

C. Moir
University of Warwick, Coventry, United Kingdom

M. Musil
Brno University of Technology, Brno, Czech Republic

P. Musil
Brno University of Technology, Brno, Czech Republic

I.G. Olaizola
Vicomtech-IK4, San Sebastián, Spain

R.R. Orozco
University of Girona, Girona, Spain

T. Ozan Aydın
Disney Research, Zürich, Switzerland

O.T. Tursun
Max Planck Institute for Informatics, Saarland, Germany

J. Unger
Linköping University, Linköping, Sweden

P. Zemčík
Brno University of Technology, Brno, Czech Republic

EDITORS' BIOGRAPHY

Alan Chalmers is professor of Visualisation and Royal Society Industrial Fellow at University of Warwick and co-founder of the spinout company, TrueDR Ltd. Previously, he was a founder and CEO of the spinout company, goHDR Ltd. He has published over 230 papers in journals and international conferences on high-fidelity virtual environments and HDR imaging, and successfully supervised 37 Ph.D. students. He is honorary president of Afrigraph, a fellow of the ERA Foundation, and formerly vice president of ACM SIGGRAPH. Together with SpheronVR, he was instrumental in the development of the world's first HDR video camera, which was completed in July 2009. From 2011 to 2015 he was chair of the EU COST Action IC1005 "*HDRi: The digital capture, storage, transmission, and display of real-world lighting*" http://ic1005-hdri.inesctec.pt/. This coordinated research and development in HDR across 25 EU countries and 44 institutions. In addition, Chalmers is a UK representative on IST/37, considering HDR standards within MPEG.

Peter Shirley is a distinguished research scientist at NVIDIA. He was formally a cofounder of two software companies and a professor at various universities. He received a B.S. in Physics from Reed College in 1985 and a Ph.D. in Computer Science from University of Illinois in 1991. He is the coauthor of several books on computer graphics and a variety of technical articles. His professional interests include interactive and HDR imaging, computational photography, realistic rendering, statistical computing, visualization, and immersive environments.

Patrizio Campisi received the Ph.D. degree in Electrical Engineering from Roma Tre University, Rome, Italy, where he is full professor at the Section of Applied Electronics, Department of Engineering. His research interests are in the area of secure multimedia communications and biometrics. Specifically, he has been working on secure biometric recognition, digital watermarking, image deconvolution, image analysis, stereo image and video processing, blind equalization of data signals, and secure communications. He has been the general chair of the seventh IEEE "Workshop on Information Forensics and Security," WIFS 2015, November 2015, Rome, Italy, and of the 12th ACM Workshop on Multimedia and Security, September 2010, Italy. He has been technical co-chair of the 1st ACM Workshop on Information Hiding and Multimedia Security, June 2013, France and of the "Fourth IEEE Workshop on Information Forensics and Security," WIFS 2012, December 2012, Spain.

He is the editor of the book "Security and Privacy in Biometrics," Springer, July 2013. He is co-editor of the book "Blind Image Deconvolution: theory and applications," CRC Press, May 2007. He is co-recipient of an IEEE ICIP06 and IEEE BTAS 2008 best student paper award and of an IEEE Biometric Symposium 2007 best paper award. He has been associate editor of IEEE Signal Processing Letters and of IEEE Transactions on Information Forensics and Security. He is currently senior associate editor of IEEE Signal Processing Letters. He is IEEE SPS Director Student Services. He is a member of the IEEE Technical Committee on Information Assurance & Intelligent Multimedia-Mobile Communications, System, Man, and Cybernetics Society, and was a member of the IEEE Certified Biometric Program (CBP) Learning System Committee.

Igor G. Olaizola obtained his degree in Electronics Engineering at the Universidad of Navarra (Tecnun) in (2001). In 2001 he worked as a research assistant in Fraunhofer Institut für Integrierte Schaltungen (IIS), Erlangen (Germany) were he worked in the development of MPEG audio decoders (MP3 & AAC) on FPGAs and DSPs.

In 2002 he started as part of the research team of Vicomtech. In 2006 he worked for Vilau (a media engineering company) in consulting, development, and deployment activities, especially in the Digital TV market. Since 2007, Igor is the head of the Digital Media department in Vicomtech where he has actively participated in R&D projects related to Media Production and Broadcasting technologies.

In 2013 he received his Ph.D. in Computer Science and Artificial Intelligence by the Faculty of Computer Science of the University of the Basque Country.

Igor participated in the COST Action IC1005 HDRi as dissemination chair and currently is associated lecturer in Tecnun. His main research interests are related to multimedia processing and analysis techniques, with a special focus on broadcasting technologies.

ACKNOWLEDGMENTS

We are most grateful to COST (European Cooperation in Science and Technology) for funding COST Action IC1005: "HDRi: The digital capture, storage, transmission, and display of real-world lighting." In particular, we would also like to thank our Science Officer, Dr. Ralph Stübner and Administrative Officer Ms. Aranzazu Sanchez for their highly valuable support, which helped ensure IC1005 was a success and which has led to a vibrant HDR community across Europe and resulted in Europe now playing a leading role in the current and future developments of HDR.

OVERVIEW OF HDR VIDEO

A. Chalmers*, P. Shirley†, D. Toffoli‡
*University of Warwick, Coventry, United Kingdom
†NVIDIA, Salt Lake City, UT, United States
‡SIM2 BV International Srl, Pordenone, Italy

Contents

1. INTRODUCTION

Welcome to the book "HDR video: Concepts, Technologies and Applications." This book contains chapters from experts in high dynamic range (HDR) video covering a wide collection of their latest research and development across all aspects of the HDR pipeline from Capture, to Manipulate and Display. Why you should read this book is that many people, the editors included, think HDR is going to be the biggest qualitative change in TV quality since color. To anyone who has seen high quality HDR content on state-of-the-art HDR displays, the step change from existing display technology is clear.

But what is HDR? It is imagery that is as subtle, bright, and vibrant as the real world. There were analog HDR technologies in the past (e.g., a good 35-mm slide with a very bright projector), but now that almost all content is digital, we can make a more precise definition: HDR is more brightness than we are used to with traditional imaging. This means "enough" bits for brightness (not the 8 bits per channel of previous technologies), and for the highest brightness of the displayed imagery to be "bright enough." What is "enough" is one of the main debates of HDR, and a source of disagreement between academia and industry. The first research papers on HDR appeared in the early 1990s (e.g., [1, 2]), so why has it taken more than 20 years for HDR to appear in consumer televisions?

First, remember we saw the same slow progress for HD TV that also took decades for the industry to deploy for the good reason that adapting an entire industry of capture, transmission, and display is a very hard process. Second, advances in LED technology were needed to make the displays commercially attractive; we have finally reached that point. Finally, the much easier transition in resolution had to be played out to its limit. We seem to have reached that in 4K, except for specialized applications. As televisions make the move toward HDR, this book also includes a chapter from an economist, Christopher Moir, that examines factors that may influence the widespread uptake of HDR (Fig. 1).

HDR is coming, and it going to impact every aspect of imaging.

This book is a major outcome of the EU Cost Action IC1005 that ran from May 2011 to May 2015 [3]. The key goals of IC1005 were to coordinate research and development in the field of High Dynamic Range (HDR) imaging across Europe, develop a vibrant community, together propose a new set of standards for the complete HDR pipeline, and establish Europe firmly as the world leader in HDR. Back in 2011, HDR was relatively unknown and the term was easily confused with HD = High Definition, that is, a screen resolution of 1920×1080 pixels. In fact, at that time, IC1005 considered supporting the term "True Brightness" to replace HDR to help minimize the confusion.

In 2012 the specification for UHDTV, ITU-R Recommendation BT.2020 was announced [4]. This contained five components to move televisions a significant step forward: Higher screen resolution 4K (3840 × 2160 pixels) and 8K (7680 × 4320 pixels); higher dynamic range; wider color gamut and higher frame rate. Consumer 4K televisions were quick to

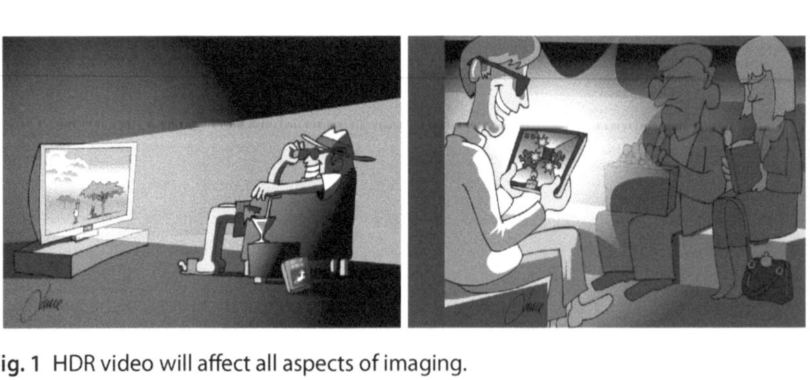

Fig. 1 HDR video will affect all aspects of imaging.

appear, but despite the significant increase in screen resolution, they failed to gain substantial market penetration. A key problem is that, unless one is very close to the screen, it is quite difficult for a viewer to see a difference between an HD and a 4K image. Of all the components of BT.2020, many user studies, for example, those by the 4Ever project [5], have clearly shown that dynamic range is the feature that users most notice, especially in a dark viewing environment. In 2016, the term HDR was thus seized upon by TV manufacturers as a "marketing tool" to sell more televisions.

This chapter documents the rise of HDR video and shows how IC1005 has tried to guide the process.

2. COST ACTION IC1005

COST Actions are European Union funded initiatives to bring together researchers, engineers, and scholars from EU and some other countries to develop new ideas and joint projects [6]. COST does not fund research itself, but rather networking, including short term scientific missions (STSMs) to enable young researchers to spend time in other labs across Europe.

Built around the three guiding principles of "*coordination*, *collaboration*, and *acceleration*," COST Action IC1005 "HDRi: The digital capture, storage, transmission, and display of real-world lighting," involved the key people and institutions undertaking or interested in HDR research and development across Europe (Fig. 2). At the end of the Action there were attendees from 44 institutions across 26 countries. Members of IC1005 played a leading role in substantially raising public awareness of HDR through their dissemination activities, including presenting the world's first end-to-end real-time HDR systems at events such as IBC (Europe's largest broadcast event) and NAB (which annually attracts over 92,000 participants), providing their vision of the future of television at the European Broadcasting Union's annual conference, and hosting a dedicated event to showcase the benefits of HDR technology to national broadcasters. From the beginning, IC1005 engaged the services of professional cartoonists Lance Bell and Danny Burgess to graphic record all the meetings. As the cartoons in this chapter show, this resulted in highly valuable material for dissemination activities and other publicity material.

A first task of the COST Action was to define (Fig. 3) the terminology used within the HDR community to ensure seamless communication between academic and industrial members. The dynamic range of a scene is the ratio of the maximum light intensity to the minimum light intensity.

Fig. 2 Guiding principles of IC1005.

Fig. 3 Defining HDR terminology.

However, what constituted *high* dynamic range was less clear. IC1005 agreed in September 2013 to use the term f-stop (or stop) to refer to the following contrast ratios:

$$\text{X f-stops} = \text{difference of } 2^X = 2^X : 1$$

So

$$\text{16 f-stops} = \text{difference of } 2^{16} = 65,536 : 1$$

This is normally noted as 100,000:1; approximately what the eye can see in a scene with no adaptation.

$$\text{20 f-stops} = \text{difference of } 2^{20} = 1,048,576 : 1$$

This is normally noted as 1,000,000:1; approximately what the eye can see in a scene with minimal (no noticeable) adaptation.

From this the following were defined:

Standard dynamic range (SDR) is ≤ 10 f-stops (aka low dynamic rage (LDR));

Enhanced dynamic range (EDR) is > 10 f-stops and ≤ 16 f-stops;

High dynamic range (HDR) is > 16 f-stops.

This definition of SDR, EDR, and HDR was subsequently adopted by the MPEG adhoc committee on HDR in their Call for Evidence document in February 2015 [7] and has largely been accepted by the scientific community. However, it is important to point out that the media industry does not make any distinction between EDR and HDR. Therefore many current commercial HDR claims are actually EDR, according to the IC1005 and MPEG scales of dynamic range. There are a number of other ways of defining dynamic range. Table 2 provides a useful means for converting from one definition to another.

One of the key impacts of IC1005, was the push for standards (Fig. 4) that can bring about the seamless connection of the three components of the end-to-end HDR pipeline: *Capture*, *Manipulate*, and *Display*. Only through such standardization will the uptake of HDR be widespread and sustainable. In this endeavor, IC1005 offered their own solution and made significant input to MPEG's goal of introducing HDR into their standards. However, MPEG chose not accept IC1005's solution and have preferred to

Fig. 4 The need for standards in the HDR pipeline.

go for SMPTE ST2084 [8] and ARIB STD-B67 [9]. Details of IC1005's proposal, together with source code, etc., are available from Ref. [10].

In addition to proposing a new standard for the entire HDR pipeline, IC1005 also delivered, via four Working Groups, the following:

- HDR file formats for individual frames used in high fidelity applications, and for complete sequences of HDR footage for distribution and delivery.
- HDR quality metrics for three different purposes: the first for assessing the quality of the image that is captured and stored, compared to the real world scene. The second to enable quality of the compressed footage to be judged, and the third metric for allowing a detailed comparison of the image that is finally displayed to the viewer compared to the real scene that was captured.
- An HDR compression benchmark: a straightforward compression benchmark, including challenging test footage, against which future commercial compression algorithms can be compared.
- An uptake plan that served to highlight at all stages of the project where the partners can make major contributions in support of the widespread adoption of HDR and generate substantial interest in the Action and HDR.

The Action also resulted in significant dissemination of information and contributed to advancing the scientific/technological state-of-the-art in the area of HDR imaging. This was achieved via scientific publications, numerous meetings, training schools, STSMs, conferences, workshops, newsletters, etc. In total, members of IC1005 chaired 4 special sessions on HDR at major international conferences; edited 3 special issues on HDR for three leading journals, and edited 2 books on HDR video, including this one.

2.1 Success Stories

In the four years of operation, COST Action IC1005 achieved significant success in three key areas:

- closing the gap in expertise,
- transfer of knowledge, and
- raising awareness of HDR.

While a number of participants in IC1005 were already world-leading authorities on HDR, many others only had an interest in the field. Access to the experts and also novel HDR equipment played an important role in

Fig. 5 IC1005 built an HDR community in Europe.

allowing these initial non-experts to become authorities in their own right and thus help to significantly raise the profile of HDR in their respective countries. The vibrant HDR community that IC1005 fostered allowed knowledge to be transferred from the academic partners to industry and the industrial needs to be transferred to the academics, to the mutual benefit of both. Now HDR is perhaps the leading topic within the television industry and amongst related standardization bodies. Below are a few examples of the successes that IC1005 has helped to bring about (Fig. 5).

2.1.1 HDR in Dental X-ray Imaging

Zeljen Trpovski, Faculty of Technical Sciences, Novi Sad,Serbia

Although Zeljen Trpovski had an interest in HDR imaging for 10 years, IC1005 spurred him on to undertake some new challenging applications. The application he chose was the use of HDR in dental X-ray images (Fig. 6). Dental radiologic images contain many details and their good distinguishing and accurate location are of significant interest in several medical fields. IC1005 provided him with the opportunity to meet with experts and gain new and interesting ideas for the continuation of his research. Particularly useful was the provision of the HDR MATLAB toolbox by Dr Banterle IT MC substitute. Preliminary results were presented at the

Fig. 6 HDR imagery for dentistry.

first International Conference and SME Workshop on HDR imaging in 2013 [11].

2.1.2 HDR Tone Mapping in Game Development

Igor Nedelkovski, University "St. Kliment Ohridski," Bitola, FYR Macedonia

Prior to COST Action IC1005 Action, researchers from the Faculty of Technical Sciences, University "St. Kliment Ohridski"—Bitola (FTS) and Motion Universe (MU), a small company owned by ex-students of the FTS were very experienced with computer graphics, but had little or no knowledge about HDR. As a result of their active participation in the COST Action, including attending the working group and management committee meetings and the Training Schools, substantial knowledge was gained; so much so that they decided to apply this knowledge to the development of a computer game for mobile platforms (Android, iOS, Blackberry). The game, Escape Medieval, was thus rendered almost entirely using HDR tone mapping (Fig. 7). As the images show, the quality of the resultant game was significantly higher than similar, competitive mobile games available on the market today. As a result of this success HDR will be introduced into computer graphics courses at FTS and the knowledge disseminated to other Universities and businesses in the Western Balkans.

Fig. 7 Tone mapped HDR images from Escape Medieval.

2.1.3 New PhD Topic at Vienna University of Technology

Margrit Gelautz, TU Vienna

The COST Action motivated TU Vienna to combine their previous experience on stereo analysis with HDR topics (Fig. 8). A new Ph.D. topic on "HDR Stereo Matching" was formulated and financially supported by the Vienna Ph.D. School of Informatics (http://www.informatik.tuwien.ac.at/teaching/phdschool). TU Vienna published papers at the two HDRi Workshops and the Special Session on HDRi at Eusipco 2014. The knowledgeable and very detailed feedback on the papers by the reviewers (chosen amongst the experts of the COST Action) gave TU Vienna valuable input to improve the papers and stimulate future work.

2.1.4 The Growth of HDR in France

Kadi Bouatouch and Rémi Cozot, University of Rennes, Ronan Boitard, Technicolor

Inspired by the vision of HDR that IC1005 was putting forward, the following companies and institutions in France (Fig. 9): Technicolor, Telecom

Fig. 8 TU Vienna Ph.D. School of Informatics has introduced a new HDR related topic.

ParisTech, University of Nantes, Dxo Lab, Dxo Sig, Binocle, Transvideo, Thomson Video Networks, AcceptTV, TF1 and Polymorph, put together the Nevex proposal. This nationally-funded project considered the design of a complete HDR chain: acquisition, compression, transmission, and display. Members of Nevex were very active in IC1005. A key output of this project was the groundbreaking Ph.D. by R. Boitard on HDR video tone mapping.

2.1.5 HDR Imaging of Rocket Launches

Brian Karr, Kennedy Space Center, Alan Chalmers, University of Warwick

Contact was made between the head of the Advanced Imaging Lab at Kennedy Space Center, Brian Karr, and the Chair of IC1005, Alan Chalmers, as a direct result of IC1005's dissemination activity at NAB in April 2013. The outcome of this meeting was that a team from goHDR/University of Warwick went to the Kennedy Space Center in July 2013 to film a rocket launch in HDR (Fig. 10).

2.1.6 World's First Complete Real-Time HDR Broadcast

Alan Chalmers, University of Warwick, Igor Olaizola,Vicomtech, Domenico Toffoli, SIM2

Assisted by earlier STSM visits between the University of Warwick, UK and Vicomtech, ES, in April 2014 a world first end-to-end live HDR pipeline was demonstrated at NAB 2014 (Fig. 11). This annual event attracts over 92,000 participants and thus it was deemed a highly appropriate venue to show this system for the first time. The system was shown on a stand at the Futures Park section of NAB as part of an IC1005 dissemination activity. The system attracted significant interest. The work for NAB 2014 led directly to the world's first professional end-to-end HDR pipeline when

Fig. 9 Frame from temporally coherent tone mapped HDR video.

Fig. 10 Capturing a rocket launch with HDR video at the Kennedy Space Center.

Fig. 11 Professional end-to-end HDR pipeline at NAB in 2015.

Fig. 12 The use of HDR in performance.

the prestigious German camera manufacturer ARRI joined the project. This professional system was showcased at NAB 2015, again as part of an IC1005 dissemination activity.

2.1.7 Evaluation of HDR Usage for Performance Art Projects

Emmanouela Vogiatzaki, RFSAT Ltd and University of Peloponnesus, Artur Krukowski, Intracom S. A. Telecom Solutions, Alan Chalmers, University of Warwick

Following a visit by Vogiatzaki to the University of Warwick in July 2014, a range of trials of HDR cameras has been undertaken to capture scenographies of performance projects in theatrical environments (Fig. 12).

The challenge was to achieve sufficient dynamic range for capturing both subjects in deep shadows simultaneously with brightly lit main subjects. The results have shown some clear advantages of using HDR technology in the performing arts, however more work is needed before such technology can be used in live performances.

3. THE HDR VIDEO PIPELINE

With HDR video, the full range of lighting in the scene can be captured and delivered to the viewer. In order to achieve such a "scene referred" ability, 32 bit IEEE floating point values are needed to be used to represent each color channel. With 96 bits per pixel (bpp), compared with just 24 bpp for SDR, this means a single HDR frame of uncompressed 4K UHD resolution (3840×2160 pixels) 4:4:4 requires approximately 94.92 MB of storage, and a minute of data at 30 fps needs 166 GB. Such a large amount of data cannot be handled on existing ICT infrastructure, and thus efficient compression is one of the keys necessary for the success of HDR video.

3.1 Capture

Until single sensors are capable of capturing the full range of light in a scene, a number of other approaches are necessary. One of the most popular methods of HDR capture is to use multiple exposures with different exposure times to create a single HDR frame [12]. However, if anything in the scene, or the camera moves, while the exposures are being taken, undesirable ghosting artefacts can occur. In this case, sophisticated deghosting algorithms, such as those described in Chapter 1, are required. This is even more of a challenge if live broadcast of HDR video is desired [13]. Another approach is to use multiple sensors with the same integration time through a single lens. A number of prototypes of such systems have been built, for example, [14] (which was able to achieve 20 f-stops at 30fps), and [15], but as yet, these are not widely available. A survey of 18 commercial cameras was undertaken in [16] to objectively measure the dynamic range they were capable of capturing. The goal was to determine their suitability for capturing all the detail during a rocket launch. Table 1 shows some results from that survey. Details of how the dynamic range of the cameras were measured is given in Chapter 4.

3.2 Compress

As mentioned above, compression is one of the keys to the successful uptake of HDR video on existing ICT infrastructure. A number of HDR

Table 1 Objective measure of camera dynamic range [16]

Camera	Stops @ 0.5 RMS noise	Vendor spec.
RED DRAGON—HDR x6, Log Film, Total	11.8 + ~3	16.5+
ARRI Alexa—Log	13.9	14.0
Canon 5DM3—Magic Lantern H.264 (ISO 400/1600)	11.4	10.0
Black Magic Cinema 4K—Film	9.0	12.0

video compression methods have been proposed in the last decade. Initially the evaluation of these was restricted by the lack of available HDR video footage. This is no longer a problem, with a number of sources of HDR video now available, including from the University of Stuttgart [17] and for the MPEG community.

As discussed further in Chapter 8, HDR video compression may be classified as *one-stream* or *two-stream* [18]. The one-stream approach utilizes a single layer transfer function to map the HDR content to a fixed number of bits, typically 10 or 12. Examples include [19–24]. One problem with one-stream methods is that they do not work for 8 bit devices, which includes a large number of legacy displays and mobile devices. With mobile devices, this is a major limitation as more and more video is being consumed on them; more than 51% of traffic on mobile devices is video viewing [25]

In two-stream compression methods, there is one input HDR video stream to the encoder that produces two bit streams as output. Two stream methods can be used for 8 bit and higher infrastructures. These streams can consist of (1) a standard compliant bit stream, for example, HEVC Main 10, H.264, etc., and (2) one another stream corresponding to additional data to reconstruct the HDR video. At the decoder, these two streams are recombined to produce the HDR video stream. Examples of two-stream compression for HDR video include [26–29].

3.3 Display

In January 2016 a consortium of TV manufacturers, broadcasters, and content producers was formed. Known as the UHD Alliance [30], their goal is to promote the criteria by which they would judge future platforms to be suitable for delivering a "premium 4K experience." Despite IC1005's definition of HDR being adopted by MPEG [7], UHD Alliance proposed two of their own definitions of HDR, both of which are "display referred":

1. 1000 nits peak brightness and <0.05 nits black level (contrast ratio 20,000:1, 14.3 f-stops),

or

2. >540 nits brightness and <0.0005 nits black level (assuming it is possible to measure 0.0005 nits, this give a contrast ratio 1,080,000:1, ≈ 20 f-stops).

These two definitions for HDR were chosen to satisfy manufacturers of different television technologies. Option 1 is for those with LED based TVs which have higher brightness, but inferior black levels, while Option 2 is for OLED based TVs which have deep blacks, but much lower peak brightness levels [31].

The UHD Alliance's "displayed referred" definition of HDR has already led to a number TV manufacturers marketing their televisions as "HDR displays." Such "consumer HDR" displays, however, have a major limitation. They do not provide an overall brighter image. Instead the average picture level (APL) of the frames is maintained and an HDR effect is achieved by providing (a) more details in the darker regions of the scene and (b) having (a few) brighter highlights [32]. Because of the HDR metadata, the TV has to be driven at its maximum backlight capacity in order to provide the required highlights. However, this means that the overall brightness of the screen cannot simply be increased as you can with an SDR display. HDR content on these displays should thus not be viewed in ambient lighting levels exceeding 5 nits [33]. The situation is even worse for HDR displays based on OLED technology. For such displays, darker than 5 nits viewing conditions, and even the absence of white walls is recommended [33]. A key reason for this limitation is that every value of the ST.2084 transfer function is directly mapped to the same specific output luminance level on all HDR displays. This absolute mapping is not flexible and is unable to adapt to different ambient lighting levels.

The 1000nit peak luminance of consumer HDR displays is significantly lower than the 4000nit HDR displays that have been available from innovative Italian company, SIM2, since 2009. Derived from the pioneering HDR displays developed by Brightside [34], SIM2's latest HDR47 display has a peak luminance of 6000 nits (Fig. 13). This is achieved via a backlight of 2202 LEDs each of which has 12 bits of brightness resolution. A key feature of the SIM2 display is the attention to high power management and heat cooling. This allows the 6000nit display to have a 1.5-kW power supply limit with a standard 110V plug. A 10,000nit display is due to be showcased at IBC in September 2016.

Fig. 13 The innovation of a 6000nit HDR display.

4. DISCUSSION

Public awareness of HDR has come a long way in the last 5 years. Most of the major television manufacturers are marketing HDR-TVs, while camera manufactures are providing more f-stops from their cameras. For compression, MPEG are also close to proposing a standard which is likely to include both the Perceptual Quantization (PQ) and Hybrid Log-Gamma (HLG) methods. In July 2016 the European Broadcasting Union (EBU) announced ITU-R BT.2100-0; their recommendations for "Image parameter values for high dynamic range television for use in production and international programme exchange" [35]. The recommendation also covers PQ and HLG. Interestingly, BT.2100 acknowledges: "The Administrations of France and the Netherlands have expressed concerns regarding the characteristics and performance of HDR-TV. Further studies are needed and may lead to a revision of this Recommendation, as appropriate, under the terms of Resolution ITU-R 1-7" [35].

There is growing evidence that the methods being considered and adopted by the manufacturers and standards bodies are not the best. PQ and HLG are computationally expensive:

$$\mathrm{PQ}_{\mathrm{encode}}(V) = \left(\frac{V^{1/m_2} - c_1}{c_2 - c_3 V^{-1/m_2}} \right)^{1/m_1},$$

where $m_1 = \dfrac{2610}{4096} \times \frac{1}{4}$, $m_2 = \dfrac{2523}{4096} \times 128$,

$$c_1 = \frac{3424}{4096}, c_2 = \frac{2413}{4096} \times 32, c_3 = \frac{2392}{4096} \times 32,$$

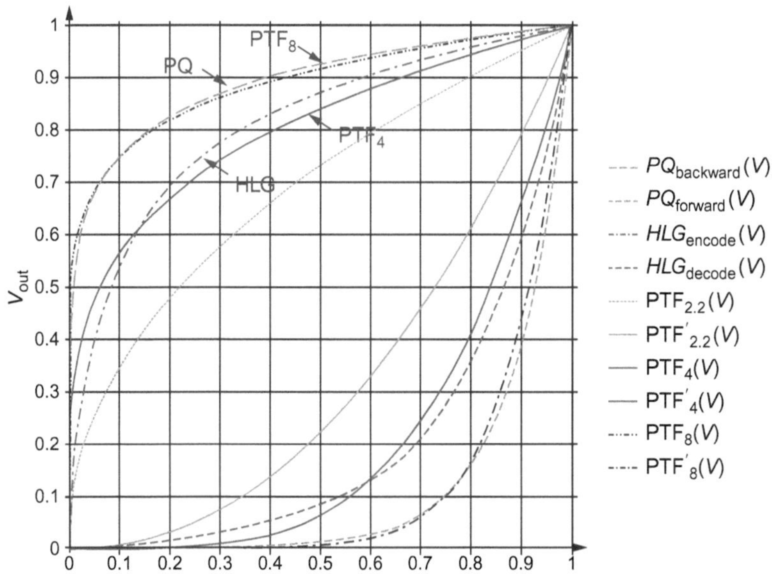

Fig. 14 Similarity of transfer function curves.

Furthermore, HLG includes a branch instruction, which limits its efficient implementation in hardware.

$$\mathrm{HLG}_{\mathrm{encode}}(V) = \begin{cases} r\sqrt{V} & V \leq 1, \\ a\log(V-b)+c & V > 1 \end{cases},$$

where $r = 0.5, a = 0.17883277, b = 0.28466892, c = 0.55991073$.

The recently proposed power transfer function (PTF) [36], on the other hand, is highly efficient (29 times more efficient than PQ, and 1.5 times more efficient than even an LUT implementation of PQ or HLG, without the memory overheads) and has been shown to deliver better quality than either PQ or HLG [36]:

$$\mathrm{PTF}_{\gamma}(V) = V^{1/\gamma},$$

where γ is a variable, for example, 4 or 8, V is normalized HDR.

As Fig. 14 shows, PTF with $\gamma = 8$ is very similar to the PQ curve, while $\gamma = 4$ closely approximates HLG. When compressing HDR video with PTF the input HDR frames first have to be normalized to the range [0, 1] with a normalization factor ℵ using the relation L = S = ℵ where S is full range HDR data. Unlike 10-bit PQ, which is limited by a peak luminance of 4000 nits (12-bit PQ has a limit of 10,000nit), PTF is not limited by a specific peak luminance.

HDR first wave—consumer HDR: The current trend of HDR, such as espoused by the UHD Alliance can be considered a "first wave" of HDR products. The HDR pipeline is very much "display referred"; constrained by the peak luminance that consumer HDR displays are capable of showing, typically 1000 nits. One positive outcome of this "first wave" is far wider user recognition of the term HDR (albeit not the definition proposed by IC1005 and adopted by MPEG). Although HDR TVs will be sold in increasing numbers in the coming months, especially as more HDR content becomes available, at some point viewers will no longer want to watch HDR in a dark room; this will give rise to a "second wave" of HDR technology.

HDR second wave—future-proof HDR: The second wave of HDR technology will be "scene referred," with the HDR pipeline no longer constrained by what a display is capable of showing. Rather the full range of lighting captured in a scene will be transmitted along the pipeline and the best image possible delivered for a given display, the ambient lighting conditions, and any creative intent. Known as true-HDR, this approach enables HDR content to be displayed directly on an HDR display while, on an SDR display, a tone mapper can be chosen, dynamically, to best suit the current scene, creative intent and ambient light conditions [18]. True-HDR enables every pixel of each HDR frame to be modulated if necessary. This allows objects to be tracked, for example, a golf ball up against a sky, of something coming off a vehicle during a rocket launch, by having their pixels at one at one exposure, while the rest of the scene could be at another. The ability to modulate individual pixels can be taken further. "Personalized pixels" can provide content creators additional creativity by enabling them to deliberately hide detail, such as clues, in areas of the scene which can only be discovered by the user interacting with the content by exploring the exposures of those pixels (Fig. 15).

This is an exciting time for HDR video: The limitations of the "first wave" will begin to be appreciated and we will move to much more flexible, compelling, future-proof "second wave." This book contains many valuable contributions that will provide the foundation for this "second wave."

APPENDIX

Table 2 shows the dynamic range conversion chart.

Table 2 Dynamic Range Conversion Chart

Dynamic Range Representations					Number of Bits Required			Terms	HVS
Density	Decades	Contrast Ratio	DR (dB)	DR (Stops)	Linear	Log	LogLuv		
Log10(CR)	10	10 Density	20Log10(CR)	Log2(CR)	1% RS	1% RS	0.27% RS		
0	10^0	1.0	0	0.0	0.0	0.0	0.0	LDR	HVS with no adaption
0.15		1.4	3	0.5	7.1	5.1	7.0		
0.2		1.6	4	0.7	7.3	5.5	7.4		
0.3		2.0	6	1.0	7.6	6.1	8.0		
0.4		2.5	8	1.3	8.0	6.5	8.4		
0.6		4.0	12	2.0	8.6	7.1	9.0		
0.8		6.3	16	2.7	9.3	7.5	9.4		
1	10^1	10	20	3.3	10.0	7.9	9.7		
1.2		16	24	4.0	10.6	8.1	10.0		
1.4		25	28	4.7	11.3	8.3	10.2		
1.6		40	32	5.3	12.0	8.5	10.4		
1.8		63	36	6.0	12.6	8.7	10.6		
2	10^2	100	40	6.6	13.3	8.9	10.7		
2.2		158	44	7.3	14.0	9.0	10.9		
2.4		251	48	8.0	14.6	9.1	11.0		
2.6		398	52	8.6	15.3	9.2	11.1		
2.8		631	56	9.3	15.9	9.3	11.2		
3	10^3	1000	60	10.0	16.6	9.4	11.3		
3.2		1585	64	10.6	17.3	9.5	11.4	EDR	
3.4		2512	68	11.3	17.9	9.6	11.5		
3.6		3981	72	12.0	18.6	9.7	11.6		
3.8		6310	76	12.6	19.3	9.8	11.7		HVS with minimal adaption
4	10^4	10,000	80	13.3	19.9	9.9	11.7		
4.2		15,849	84	14.0	20.6	9.9	11.8		
4.4		25,119	88	14.6	21.3	10.0	11.9		
4.6		39,811	92	15.3	21.9	10.1	11.9		
4.8		63,096	96	15.9	22.6	10.1	12.0		

Continued

Table 2 Dynamic Range Conversion Chart—cont'd

Dynamic Range Representations					Number of Bits Required			Terms	HVS
Density	**Decades**	**Contrast Ratio**	**DR (dB)**	**DR (Stops)**	**Linear**	**Log**	**LogLuv**		
Log10(CR)	**10**	**10 Density**	**20Log10(CR)**	**Log2(CR)**	**1% RS**	**1% RS**	**0.27% RS**		
5	10^5	100,000	100	16.6	23.3	10.2	12.1	HDR	
5.2		158,489	104	17.3	23.9	10.2	12.1		
5.4		251,189	108	17.9	24.6	10.3	12.2		
5.6		398,107	112	18.6	25.2	10.3	12.2		
5.8		630,957	116	19.3	25.9	10.4	12.3		
6	10^6	10^6	120	19.9	26.6	10.4	12.3		
6.2		$10^{6.2}$	124	20.6	27.2	10.5	12.4		HVS with full adaption
6.4		$10^{6.4}$	128	21.3	27.9	10.5	12.4		
6.6		$10^{6.6}$	132	21.9	28.6	10.6	12.5		
6.8		$10^{6.8}$	136	22.6	29.2	10.6	12.5		
7	10^7	10^7	140	23.3	29.9	10.7	12.5		
7.2		$10^{7.2}$	144	23.9	30.6	10.7	12.6		
7.4		$10^{7.4}$	148	24.6	31.2	10.7	12.6		
7.6		$10^{7.6}$	152	25.2	31.9	10.8	12.7		
7.8		$10^{7.8}$	156	25.9	32.6	10.8	12.7		
8	10^8	10^8	160	26.6	33.2	10.9	12.7		
9	10^9	10^9	180	29.9	36.5	11.0	12.9		
10	10^{10}	10^{10}	200	33.2	39.9	11.2	13.1		
11	10^{11}	10^{11}	220	36.5	43.2	11.3	13.2		
12	10^{12}	10^{12}	240	39.9	46.5	11.4	13.3		
13	10^{13}	10^{13}	260	43.2	49.8	11.6	13.4		
14	10^{14}	10^{14}	280	46.5	53.2	11.7	13.5		
15	10^{15}	10^{15}	300	49.8	56.5	11.8	13.6		
16	10^{16}	10^{16}	320	53.2	59.8	11.9	13.7		
17	10^{17}	10^{17}	340	56.5	63.1	11.9	13.8		
18	10^{18}	10^{18}	360	59.8	66.4	12.0	13.9		

Courtesy of Brian Karr.

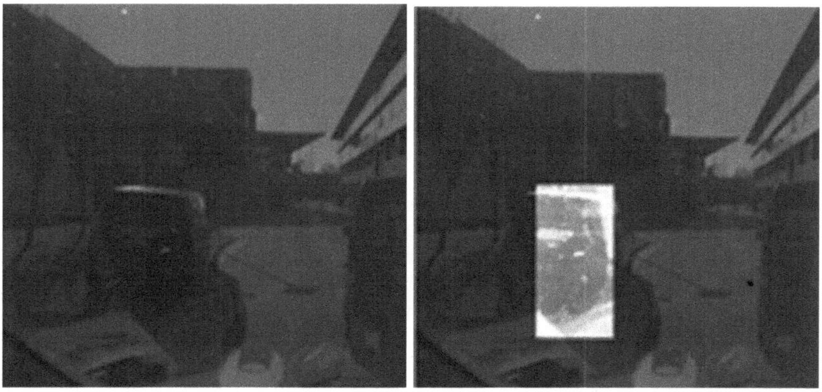

Fig. 15 Personalized pixels: The hidden character can be seen in the image on the right by exploring the pixels of the HDR image [37].

REFERENCES

[1] K. Chiu, M. Herf, P. Shirley, S. Swamy, C. Wang, Zimmerman K., Spatially Nonuniform Scaling Functions for High Contrast Images, in: Graphics Interface '93, 1993.

[2] J. Tumblin, Rushmeier H., Tone reproduction for realistic images, IEEE Comput. Graph. Appl. 13 (1993).

[3] http://www.cost.eu/COST_Actions/ict/IC1005.

[4] ITU-R BT.2246-2, The present state of ultra-high definition television, tech. rep. 2012 http://www.itu.int/dms_pub/itu-r/opb/rep/R-REP-BT.2246-2-2012-PDF-E.pdf.

[5] http://www.4ever-project.com/home/.

[6] http://www.cost.eu/.

[7] A. Luthra, E. François, W. Husak (Eds.), Requirements and Use Cases for HDR and WCG Content Distribution, ISO/IEC JTC 1/SC 29/WG 11 (MPEG) Doc. N15084, February 2015.

[8] ST 2084:2014 : High Dynamic Range Electro-Optical Transfer Function of Mastering Reference Displays, August 2014.

[9] STD-B67: Essential parameter values for the extended image dynamic range television (EIDRTV) system for programme production, tech. rep., Association of Radio Industries and Businesses (ARIB), July 2015.

[10] http://ic1005-hdri.inesctec.pt/.

[11] Ž. Trpovski, V. Kerkez, P. Vučinic, HDR processing for legibility enhancement of radiographic images, in: First International Conference and SME Workshop on HDR imaging, 2013.

[12] P.E. Debevec, J. Malik, Recovering high dynamic range radiance maps from photographs, in: SIGGRAPH '97, ACM Press, 1997.

[13] J. McNamee, J. Hatchett, K. Debattista, A.G. Chalmers, Live HDR Video Streaming on Commodity Hardware, in: SPIE 2015, August 2015.

[14] A. Chalmers, G. Bonnet, F. Banterle, P. Dubla, K. Debattista, A. Artusi, C. Moir, High Dynamic-range video solution, in: ACM SIGGRAPH ASIA, Art Gallery & Emerging Technologies, ACM, 2009.

[15] M.D. Tocci, C. Kiser, N. Tocci, P. Sen, A versatile HDR video production system, ACM Trans. Graph. 30 (4), Article 41 (July 2011).

[16] B. Karr, A.G. Chalmers, K. Debattista, High dynamic range digital imaging of spacecraft, in: F. Dufaux, P. Le Callet, R. Mantiuk, M. Mrak (Eds.), High Dynamic Range—From Acquisition to Display and Applications, Elsevier, 2015.

[17] https://hdr-2014.hdm-stuttgart.de/.

[18] A. Chalmers, J. McNamee, J. Hatchett, R. Mukherjee, I. Olaizola, K. Debattista, 12 bits is simply not enough for HDR video! 2015, BEC 2015, NAB.

[19] R. Mantiuk, G. Krawczyk, K. Myszkowski, H.-P. Seidel, Perception-motivated high dynamic range video encoding, ACM Trans. Graph. 23 (3) (2004) 733–741.

[20] A. Motra, H. Thoma, An adaptive logluv transform for high dynamic range video compression, in: 17th IEEE International Conference on Image Processing (ICIP), IEEE, 2010.

[21] Y. Zhang, E. Reinhard, D. Bull, Perception-based high dynamic range video compression with optimal bit-depth transformation, in: 18th IEEE International Conference on Image Processing (ICIP), IEEE, 2011.

[22] S. Miller, M. Nezamabadi, S. Daly, Perceptual signal coding for more efficient usage of bit codes, in: SMPTE Conference, 2012.

[23] T. Borer, Non-linear opto-electrical transfer functions for high dynamic range television, BBC White Paper, 2014.

[24] J. Hatchett, K. Debattista, R. Mukherjee, T. Bashford Rogers, Chalmers A.G., Highly efficient HDR video compression, in: ISO/IEC JTC1/SC29/WG11 MPEG2016/JCTVC-W0072, February 2016.

[25] Ooyala, Global video index q3 2014., Report, International Telecommunication Union, Silicon Valley, USA, 2014.

[26] R. Mantiuk, A. Efremov, K. Myszkowski, H.-P. Seidel, Backward compatible high dynamic range MPEG video compression, ACM Trans. Graph. 25 (3) (2006) 713–723.

[27] C. Lee, C.-S. Kim, Rate-distortion optimized compression of high dynamic range videos, in: 16th European Signal Processing Conference, 2008.

[28] HDR Video Data Compression Devices and Methods 2009, EP Patent 2144444.

[29] K. Debattista, T. Bashford-Rogers, E. Selmanovic, R. Mukharjee, A.G. Chalmers, Optimal exposure compression for high dynamic range content, Vis. Comput. Int. J. Comput. Graph. 31 (6-8) (2015) 1089–1099.

[30] www.uhdalliance.org.

[31] http://www.trustedreviews.com/opinions/ultra-hd-premium.

[32] http://www.hdtvtest.co.uk/news/4k-vs- 201604104279.htm.

[33] T. Schulte, J. Barsotti, HDR Demystified: Emerging UHDTV systems, Technical paper, 2016.

[34] H. Seetzen, W. Heidrich, W. Stuerzlinger, G. Ward, L. Whitehead, M. Trentacoste, A. Ghosh, Vorozcovs A., High dynamic range display systems, ACM Trans. Graph. 23 (3) (2004), 760–768.

[35] ITU-R. BT.2100-0, Image parameter values for high dynamic range television for use in production and international programme exchange, http://www.itu.int/dms_pubrec/itu-r/rec/bt/R-REC-BT.2100-0-201607-I!!PDF-E.pdf.

[36] J. Hatchett, K. Debattista, R. Mukherjee, T. Bashford Rogers, Chalmers A.G., An evaluation of power transfer functions for HDR video compression, The Visual Computer, to appear 2016.

[37] A.G. Chalmers, B. Karr, R. Suma, C. Moir, Debattista K., Fifty shades of HDR, in: IEEE Digital Media Industry & Academic Forum, July 2016.

PART 1

Capture

CHAPTER 1

Ghosting in HDR Video

A.O. Akyüz*, O.T. Tursun†, J. Hasić-Telalović‡, K. Karađuzović-Hadžiabdić‡
*Middle East Technical University, Ankara, Turkey
†Max Planck Institute for Informatics, Saarland, Germany
‡International University of Sarajevo, Sarajevo, Bosnia and Herzegovina

Contents

1. INTRODUCTION

High-quality HDR videos can be captured in the following two ways: (1) by using dedicated video capture hardware that has improved dynamic range and (2) by using standard hardware to capture a set of frames with

High Dynamic Range Video
http://dx.doi.org/10.1016/B978-0-12-809477-8.00001-7

alternating exposures and then combining these frames for extending the dynamic range. We will call these two groups of techniques *simultaneous* and *sequential* HDR video capture methods.

Simultaneous HDR video capture typically entails the use of sensor elements with improved dynamic range. This method is usually employed by commercial products. However, it can also be performed by spatially varying the exposure of each sensor element, spatially varying the transmittance of neutral density filters placed in front of the sensor, or using a beam-splitting system to redirect the incoming light into multiple sensors, each set to a different exposure value. In this latter case, it becomes necessary to merge the output of the individual sensors to obtain an HDR frame.

Sequential HDR video capture, on the other hand, involves capturing of frames with alternating low and high exposures. This variation can be achieved by changing the exposure time, aperture size, or the sensitivity (ISO) of the camera between frames. In these techniques, each frame of the video is captured in low dynamic range (LDR), but has the potential to be converted to HDR by using the information present in the adjacent frames. These sequential techniques have recently become popular thanks to programmable digital cameras or third party software that allows capturing such alternating sequences. For example, Magic Lantern [3] is a software that furnishes most Canon DSLR cameras with this feature.

For both simultaneous and sequential systems, the merging of multiple exposures is motivated by the fact that each exposure contains details for different parts of the scene corresponding to different light levels. By merging this information, one can obtain an HDR frame that well represents a wider range of light levels of the captured scene.

However, the merging process may itself give rise to artifacts if the merged exposures are inconsistent with each other. Such inconsistencies may occur, for example, due to camera and object motion. In the case of camera motion, the exposures can be registered by global alignment techniques. However, this can still be problematic due to the parallax effect in which nearby objects are displaced more than the distant ones. In this case, a simple global alignment would not be sufficient. If the objects themselves are dynamic, this generates more complicated motion patterns, especially if this is combined with the camera motion. The field of HDR video deghosting typically deals with problems of both kinds.

2. IMAGE ACQUISITION MODEL

In this section, we formalize the concept of creating an HDR image (or video frame) from multiple exposures. Here, we also introduce a terminology that we will use when describing the deghosting methods in more detail. Such a consistent terminology will allow focusing on the algorithmic differences rather than symbolic ones.

We assume that for frame i, each sensor element (x, y) is exposed to a radiance value $E(x, y)$ for a duration of Δt_i seconds. This results in the total collected charge of $E(x, y)\Delta t_i$ units. This collected charge passes through various transformations such as analog to digital conversion, quantization, gamma correction, and various custom transformations defined by the camera manufacturer. The net effect of these transformations is represented by using a function f, which is known as the camera response function [4]. Thus the relationship between the input irradiance and output pixel values, $Z_i(x, y)$, can be written as:

$$Z_i(x, y) = f(E(x, y)\Delta t_i). \tag{1}$$

As the goal of HDR image and video generation is to recover the scene radiance values, it is necessary to estimate the (inverse) camera response function to reverse the above relationship:

$$E(x, y) = \frac{f^{-1}(Z_i(x, y))}{\Delta t_i}. \tag{2}$$

However, due to limitations of the image sensor, not all pixels provide useful information about the scene. For example, it is impossible to estimate the correct radiance value for saturated (under- or overexposed) pixels. Furthermore, each pixel is a noisy measurement of the scene radiance. As a result, estimating the true radiance from a single exposure is generally not reliable. To address this problem, various solutions have been proposed. A simple and commonly used solution is to combine multiple exposures of the same scene captured simultaneously or in rapid succession.

Typically, multiple exposures are combined by assigning a weight to each of them based on the reliability of each exposure. First of all, if $Z_i(x, y)$ was clipped due to being under- or overexposed, this measurement does not represent the true scene radiance, hence its influence should be minimized. Second, the signal-to-noise ratio and the coarseness of quantization are generally not uniform across the valid range of the sensor. Therefore, both of these factors are taken into account to derive a final weighting function, w, that can be used to combine the sensor irradiances:

$$E(x, y) = \frac{\sum_{i=1}^{N} w(Z_i(x, y)) \frac{f^{-1}(Z_i(x, y))}{\Delta t_i}}{\sum_{i=1}^{N} w(Z_i(x, y))}. \tag{3}$$

Here, for notational simplicity, we indicated the weighting function as to take a single input, which is the pixel value. However, it should be remembered that w may actually depend on various other factors such as the exposure time, camera response function, and the assumed noise model. We refer the reader to Granados et al. for a detailed review of different ways of setting the weighting function [5].

For color images, the typical practice is to treat each color channel independently to obtain E_i^r, E_i^g, and E_i^b for red, green, and blue color channels, respectively. However, to avoid potential color casts, it may be desirable to use a combined value, such as luminance, in the computation of the weighting function [6].

This simple formulation leads to blending artifacts if the corresponding pixels in different exposures belong to different parts of the scene. This could happen due to various factors, such as the movement of the camera, objects, and changes in illumination. To address this problem, Eq. (3) must be revised such that it allows us to combine pixels at different spatial coordinates. This may be accomplished by replacing (x, y) with (u_i, v_i) on the right-hand side of Eq. (3), where (u_i, v_i) represents the coordinates of a pixel in exposure i in some neighborhood of (x, y). Ideally, all (u_i, v_i) coordinates represent the same scene point in all exposures. In general, HDR image and video deghosting methods differ with respect to various strategies they employ to find these corresponding pixels.

3. HDR IMAGE DEGHOSTING METHODS

As most HDR video deghosting methods are based on or inspired from HDR image deghosting methods, in this section we will summarize their main characteristics. For a detailed review of HDR image deghosting methods, we refer the reader to a survey paper by Tursun et al. [7].

A high-level taxonomy of HDR image deghosting methods is given in Fig. 1. According to this taxonomy, image deghosting methods are divided into four main categories as (1) global exposure registration,

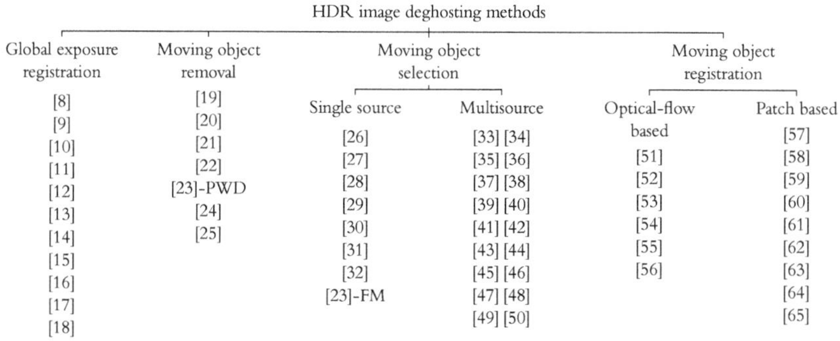

Fig. 1 Taxonomy of HDR image deghosting methods.

(2) moving object removal, (3) moving object selection, and (4) moving object registration.

The first category only involves methods that perform global registration of the multiple exposures. While some of these techniques only handle translational misalignments such as Ward [10] and Akyüz [17], others can handle both rotational and translational misalignments such as [14]. In general, these techniques assume that the captured scene is static or the moving objects occupy small regions, as not to interfere with the registration process. However, as noted in Section 1, if the captured scene has multiple objects at different distances from the camera, global misalignments caused by pure camera motion may still be problematic due to the parallax effect.

The second category involves methods that eliminate moving objects all together. In other words, they estimate the static background of the captured scene. A pioneering example of this group of algorithms is Khan et al. [19], which assumes that most pixels belong to static regions. By considering the similarity of pixels to each other in small spatiotemporal neighborhoods, the influence of dissimilar pixels (i.e., pixels belonging to moving objects) is minimized.

The third category, moving object selection, is comprised of deghosting methods that include moving objects in the output HDR image. *Single-source* methods may use only one reference exposure (e.g., the middle exposure) in all dynamic regions or they may choose a different source exposure for dynamic region based on the well-exposedness of the exposures. *Multisource* methods, on the other hand, try to combine as many exposures as possible that are consistent with each other in each dynamic region. For example, Oh et al. [42] postulate the HDR reconstruction problem as rank minimization of a matrix in which each column corresponds to an input exposure. By minimizing the rank of this matrix, the authors obtain pixels that are consistent with each other. Normally, this yields the static background of the captured scene. However, their method allows the user to choose a reference exposure which may contain moving objects of interest. The remaining exposures are then combined with this reference exposure based on how well they model the background.

The last category of HDR deghosting algorithms contains the most advanced methods in that they involve registering individual pixels across different exposures. The first subcategory is made up of optical flow based methods which work well when the motion is relatively small and there are a few occlusions (the tracked pixels do not get lost due to being hidden by another object). The second subcategory, on the other hand, describes each pixel by using a patch of varying sizes around that pixel and tries to find the best matching pixels by considering the similarity of their patches. In general, these methods appear to produce the highest quality HDR images with Sen et al. [62] being a pioneering example.

The large number of HDR image deghosting methods gave rise to various studies that evaluate their performance. Among these, both subjective and objective evaluations have been conducted. We refer the reader to Tursun et al. for a review of these methods [66].

4. HDR VIDEO CAPTURE METHODS

As discussed in the beginning of this chapter, there are various HDR video capture methods, termed *simultaneous* capture, that entirely avoid the deghosting problem. However, these methods require changes in the camera hardware and therefore cannot be readily used by most photographers and enthusiasts. These methods are discussed in this section. Alternatively, one can use existing digital cameras to capture the successive frames with different exposures, termed *sequential* capture. This results in a sequence of short and long exposures (sometimes alternation occurs between more than two exposures). In a post processing step, these exposures are registered using various techniques and then combined into HDR frames to form an HDR video. These techniques are divided into two groups based on whether they deal with ghosting at a *global* (Section 5) or *local* (Section 6)

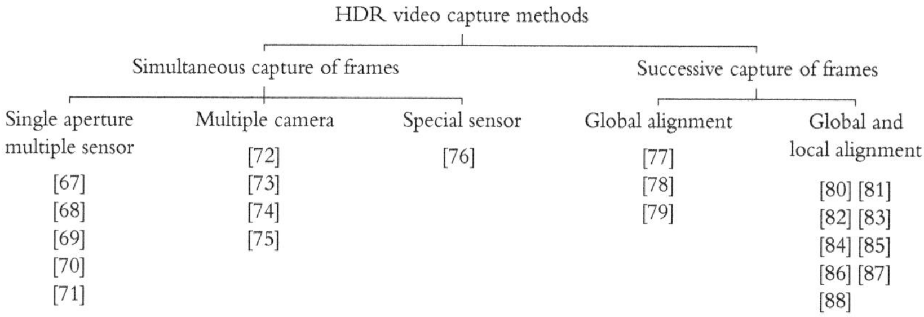

Fig. 2 Taxonomy of HDR video capture methods.

level. A taxonomy including both simultaneous and successive methods is given in Fig. 2.

4.1 Single Aperture Multiple Sensor Solutions

These groups of algorithms record the scene from a single aperture. The light that enters the aperture is then projected onto multiple sensors through a beam-splitting element. Each sensor is set to a different exposure value, which allows it to capture a different range of illumination. The sensor outputs are then combined to produce an HDR frame. This form of image generation is also called split-aperture imaging [2]. In general, this group of algorithms entirely avoids alignment and ghosting problems, as all sensors receive the same input.

A pioneering algorithm of this kind, in fact, utilizes a single sensor placed behind a transparency mask with multiple bands [89]. In other words, although there is a single sensor, different parts of the sensor receive different amounts of light depending on the transparency of the mask that is adjacent to it. The camera is designed as a panoramic camera, which rotates around its central axis. This allows the same scene point to be projected on different parts of the sensor with different exposures. As such, each scene point will be captured under multiple exposures. As the angular step size of the motor is known, one can compute the corresponding sensor positions of the same scene points, allowing them to be merged into a panoramic HDR image. This work does not propose a solution for dealing with potential changes in the scene during the capture process and therefore, it is suitable to capture only static scenes. A later work by the same authors extends this method to capture images at high frame rates, which are suitable to capture HDR videos, by using a beam-splitting mirror [67].

Wang and Raskar use a split-aperture camera to capture HDR videos [68]. The corner of a cube is used to generate a three-face pyramid, which transmits light toward three CCD sensors. Each sensor is adjacent to an ND filter with transmittances of 1, 0.5, and 0.25. The sensors are carefully aligned with respect to the pyramid such that the sensors are normal to the optical axis.

The beam-splitting approaches discussed earlier are wasteful of light in that they admit only a fraction of the light toward sensors. For example, if three-way beam-splitting is used, each sensor receives only one-third of the incident light. This light is further attenuated by the ND filters adjacent to the sensors. This reduction in light compels the photographer to use longer exposure times, which in turn may introduce problems such as thermal noise and motion blur. To address this problem, Tocci et al. propose a novel system that uses two beam-splitters which allow reusing the optical path to improve light efficiency [69]. Furthermore, they modify the HDR merging equation to consider the neighbors of a pixel rather than treating each pixel individually. This mitigates the cross-talk artifacts that may occur, for example, if a neighboring pixel is saturated, but the pixel itself not. Furthermore, it allows a smoother transition between pixels taken from different exposures.

The methods discussed earlier assume that sensors are perfectly aligned such that the same scene point is projected to the same locations in different sensors. However, this may be difficult to achieve in practice. Kronander et al. address this problem by assuming there exists an affine transform relating the coordinate system of each sensor to a virtual reference [70]. This transform is found in an offline stage by matching the corners of a checkerboard pattern. This method is later extended to capture HDR videos that perform demosaicking, denoising, alignment, and HDR assembly within a unified approach [71].

4.2 Multiple Camera Solutions

These group of algorithms utilize two or more cameras installed on a custom rig. Each camera captures the scene from its own viewpoint and with a different exposure time. The recorded frames must be registered prior to combining them to create HDR frames.

The first work that utilizes a multicamera solution was proposed by Ramachandra et al. [72]. In their system, the authors utilize a two-camera setup, whereby one camera is set to a short exposure time and the other to a long exposure time. As the longer exposed frames may be subject to motion-blur, the first step of their algorithm is deblurring of the long exposure. To this end, first, the histogram of the short exposure is matched to that of the long exposure to normalize exposure differences. Then for each 128 × 128 image patch for the blurry image (called target patch), a corresponding reference patch is found from the sharp exposure. Next, both the target and reference patches are subjected to a multiscale multiorientation pyramid decomposition. Then several features extracted from the reference subbands (such as the moments and the correlation of the coefficients) are applied as constraints on the corresponding features of the target subbands. The collapsed target subbands then produce the deblurred long exposure. However, because there could be local (smaller than the 128 × 128 patch size) variations between the patches, this process may produce ringing artifacts. These artifacts are fixed by applying a deringing filter. Once the long exposure is sharpened by applying this process and optical-flow based registration is performed, as in Kang et al. [80], the HDR frame is reconstructed using Debevec and Malik's algorithm [4].

Bonnard et al., on the other hand, propose a system composed of eight cameras [73]. These cameras are installed in a camera box with aligned and synchronized objectives [90]. Normally, this setup is designed to produce 3D content for autostereoscopic displays. The authors furnish this system with HDR capabilities by placing ND filters in front of the camera objectives. This produces eight differently exposed frames that are captured simultaneously. The disparity maps between these frames are computed by using Niquin et al.'s algorithm [91]. Before the application of this algorithm, differently exposed frames are normalized to the same average intensity level, using the light blockage ratios of the ND filters. After the disparity map calculation, each pixel is assigned to a set of matching pixels from the other views. This information is used to generate eight autostereoscopic HDR frames using the HDR reconstruction algorithm of Debevec and Malik [4].

The system proposed by Bätz et al. [75] is comprised of two cameras and two workflows. In the offline calibration workflow, the response curves of both cameras are estimated. This has to be done only once for a given pair of cameras. The main workflow is comprised of several stages. In the first one, stereo rectification is performed, which results in geometrically transformed images such that a given point in the left view resides on the

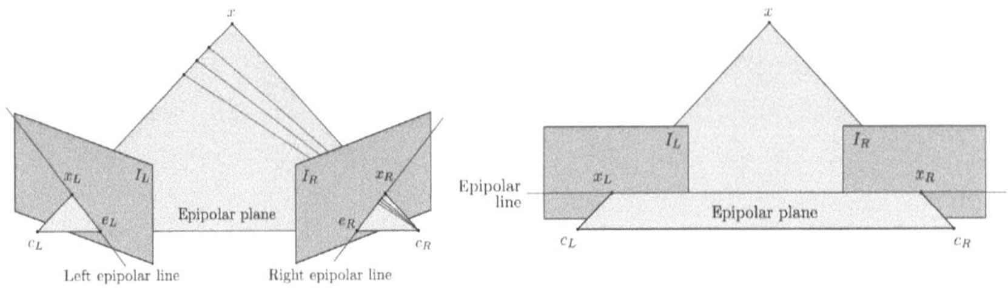

Fig. 3 The status of epipolar lines before (*left*) and after (*right*) the epipolar rectification performed by Bätz et al. [75] for a pair of pixels x_L and x_R corresponding to the same scene point x. *(Adapted from M. Bätz, T. Richter, J.-U. Garbas, A. Papst, J. Seiler, A. Kaup, High dynamic range video reconstruction from a stereo camera setup, Signal Process. Image Commun. 29 (2) (2014) 191–202.)*

same scanline as in the right view (see Fig. 3). This simplifies determining correspondences between pixel values, which are performed in the second stage. For measuring similarity between pixel values, zero mean normalized cross correlation (ZNCC) is used as it is resistant to exposure variations. The determined disparity maps are then smoothed to avoid abrupt changes between neighboring pixels.

In the third stage, the disparity map is used to warp the source exposure onto the target exposure. To this end, the target exposure is selected as the one with the fewest number of under- and over-exposed pixels. Backward warping is used which, for every pixel in the target exposure, computes a sampling position from the source exposure. As this is likely to produce noninteger sampling coordinates, bilinear interpolation is used to compute the final color value. Finally, the HDR frame is computed by merging the target exposure with the warped one. However, merging is only performed for pixels that are under- or overexposed in the target exposure, which limits the influence of potential errors that may be introduced in the earlier stages.

Similar to Bonnard et al. [73], the system proposed Selmanovic et al. aspires to generate a stereoscopic HDR video [74]. Their system is comprised of two video cameras with one camera being an HDR camera [92] and the other an LDR one. The goal is to generate the missing HDR frame by using the information from the existing HDR and LDR frames. To this end, three methods are proposed. The first one is based on warping the HDR frame toward the LDR frame based on a disparity map, similar to the method of Bätz et al. [75]. The second method involves directly expanding the dynamic range of the LDR frame using a novel inverse tone mapping approach. The third method is a hybrid of the two in which the well-exposed regions of the LDR frame are expanded and the under- and overexposed regions are warped from the HDR frame. The authors conclude that the hybrid approach produces the highest quality results.

4.3 HDR Sensor Solutions

Various algorithms discussed earlier have been necessitated due to the LDR nature of image sensors. Ideally, using an HDR image sensor would solve the problems of the aforementioned algorithms. While not being the primary focus of this chapter, here we will briefly review HDR sensor solutions for the sake of completeness.

There are several ways to design an HDR image sensor. In general, an image sensor can be considered as an array of photodetectors with each photodetector connected to a readout circuit. There may be an optional layer of optical mask in front of the photodetectors as well. An HDR sensor can be produced by either spatially varying the transmittance of the optical mask [76, 93], by modifying the readout circuitry [94], or by using HDR photodetectors [95].

The idea of using a spatially varying filter resembles that of using a Bayer color filter array to produce color images. However, instead of being selective of the wavelength, spatially varying transmittance filters transmit different amounts of light onto each photodetector. A pioneering work in this direction was proposed by Nayar and Mitsunaga [93], who used a regular array of optical filter elements aligned with photodetectors. The final HDR pixel values are obtained by computing the average of the neighboring pixel values with different exposures. However, this approach results in a slight reduction in resolution, especially in regions that contain sharp edges. Schöberl et al. show that the resolution can be improved if one uses a nonregular pattern for the optical mask due to the sparsity characteristics of natural images in the Fourier domain [76]. In a more recent work, it has been shown that high quality HDR images can be obtained from a single coded exposure using convolutional sparse coding [96].

As for modifying the readout circuitry, several alternatives have been proposed. These are (1) measuring time-to-saturation, (2) performing multiple nondestructive readouts during the capture process, (3) readout with asynchronous self-reset, and (4) synchronous self-reset with residue readout. A quantitative evaluation of these four groups of techniques was conducted by Kavusi and Gamal, who conclude that each technique has different advantages and disadvantages [94].

Finally, an HDR image sensor can be obtained by directly extending the dynamic range of the photodetecting elements. As a reference, we refer the reader to Zhou et al. for a design that involves punchthrough enhanced phototransistors [95].

5. GLOBAL VIDEO DEGHOSTING

Global deghosting algorithms are typically concerned with aligning input frames as a whole without considering local object motions. As such, they are suitable to be used if the scene is stationary or the movement is small with respect to the capture rate of the camera.

5.1 View Dependent Enhancement of the Dynamic Range of Video [77]

Among the first global approaches to HDR video deghosting problems, Niskanen presented a global alignment method that produces results that are suitable to be used by computer vision algorithms rather than to be viewed by human observers [77]. Deghosting is achieved by aligning SIFT features [97] with RANSAC sampling [98] to compensate for global motion. Therefore, this technique is only applicable when local motion between the frames is negligible.

The video is generated by combining sequential LDR frames with varying exposure times. A CCD camera was used with linear or known camera response. The acquisition process has been augmented by two techniques that control exposure times adaptively. They are based on the information content. The first technique involves maximizing the entropy of observations;

$$(\Delta t_1, \ldots, \Delta t_N) = \underset{\Delta t_1, \ldots, \Delta t_N}{\operatorname{argmax}} \; H(O_1, \ldots, O_N), \tag{4}$$

where $O_1, \ldots, O_N$ are the real scene (S) observations, $H(.)$ is the entropy of observations, and $\Delta t_1, \ldots, \Delta t_N$ are the exposure times to be selected.

The second technique maximizes the average SNR for all I pixels in the resulting HDR image:

$$(\Delta t_1, \ldots, \Delta t_N) = \underset{\Delta t_1,\ldots,\Delta t_N}{\operatorname{argmax}} \frac{\sum_{x,y} S^2(x,y)}{\sum_{x,y} \mathcal{N}^2(x,y)}, \tag{5}$$

where $S(x,y) = r\sum_{i=1}^{N} I_i(x,y)$ with $I_i(x,y)$ being the exposure time normalized and unsaturated signal level for a given pixel in the ith exposure and r is the number of photons required to gain one level of intensity. $\mathcal{N}(x,y)$, on the other hand, represents the noise level and it is modeled as a sum of the photon shot and quantization noise:

$$\mathcal{N}(x,y) = \sqrt{r\sum_{i=1}^{N} I_i + N\frac{r^2}{12}}. \tag{6}$$

Using either technique, the optimal exposure times can be computed from frame histograms. It is shown by the author that this technique can select suitable exposure times at 30 FPS among 500 different candidate values.

Once the frames are captured, they are registered using RANSAC sampled SIFT features [97]. This is followed by the construction of HDR frames that are based on modeling light radiation as a Poisson process. In the evaluation of this approach, it was noticed that for small motion, to a certain extent, using the multiple frames without alignment gives better performance than using fewer frames with motion correction.

5.2 Histogram-Based Image Registration for Real-Time HDR Videos [78]

The motivation of this method is to capture and display HDR video in real-time. However, it only considers global translational motion between the frames to be registered. This is done by creating row and column histograms that contain counts of dark and bright pixels in a row or column, and maximizing the correlation between the histograms of two consecutive frames. The two-dimensional search problem is reduced to two one-dimensional searches. The reduction in computation time enables real-time HDR video recording. The robustness of the estimation is increased by Kalman filter application [99].

The image capturing is based on the method of [100]. It achieves efficiency by doing only partial reexposures of a single frame. At first, a full resolution frame is captured at optimal exposure (shutter and gain) setting. This frame is then analyzed to indicate which regions of the frame need reexposures. Usually these regions are frame partials. Additional higher or lower exposures are then captured for these regions only, which reduces the capture and processing times. The reexposures are analyzed and, if needed, additional partial reexposures are captured. The captured images are converted to *Yxy* color space and all subsequent computations are performed on the *Y* (luminance) channel.

The process of registration creates a two-dimensional integer translation vector $\vec{v}_i$ between a parent exposure and each reexposure. Summing up all translation vectors along the path from an exposure to the base frame, one can calculate an absolute shift between each exposure and the base frame.

Median threshold bitmaps (MTBs) as described in [10] are used for estimation of translation vectors for a pair of images. As MTBs create an image that contains roughly 50% white and 50% black pixels, it produces similar bitmaps for different exposures. Once the MTBs are computed, the number of black pixels in each column is counted and these counts are collected in column histograms for each MTB.

The MTB algorithm starts with building a regular histogram with 256 bins over the brightness values of I_{i-1} and its reexposure I_i. Then the median brightness value m_i for reexposure I_i to be used as threshold is determined. Fifty percent of pixels greater than this threshold will be white, while the remaining will be black. The near median pixels are ignored for improved robustness.

The exposure values e_{i-1} and e_i of the images as well as the camera response function f are assumed to be known. To save histogram computation over I_{i-1}, m_{i-1} is calculated as follows:

$$m_{i-1} = f\left(f^{-1}(m_i)\frac{e_{i-1}}{e_i}\right). \tag{7}$$

For the purpose of this method, it is sufficient to determine the thresholds m_{i-1} and m_i, and the MTBs are actually not created.

Next, the horizontal and vertical components of the translation vector $\vec{v}_i = (x_i, y_i)$ between a parent exposure and each reexposure are computed. The horizontal (vertical) component is estimated by computing normalized cross correlation (NCC) between the two column (row) histograms.

Assuming that the width and height of image I_i are w_i and h_i, respectively, for each column $j = 1, \ldots, w_i$, the column histogram $B_i^x(j)$ of exposure I_i is the number of black pixels and is defined as:

$$B_i^x(j) = |\{I_i(j,k) < m_i - T; k = 1, \ldots, h_i\}|, \tag{8}$$

where $I_i(j,k)$ is the pixel value at position (j,k) and $|.|$ denotes set cardinality. Analogously, histogram W_i^x counts the number of white pixels. The two histograms are created for I_{i-1} in the same fashion. These four histograms are used to estimate the horizontal shift component, x_i. This is a search problem that maximizes the NCC between I_{i-1} and I_i histograms:

$$\mathrm{NCC}(s) = \frac{C}{\sqrt{N_1 N_2}}, \tag{9}$$

where C, N_1, and N_2 are computed as follows:

$$C = \sum_{j=1}^{w_i} (W_i^x(j) W_{i-1}^x(j-s) + B_i^x(j) B_{i-1}^x(j-s)), \tag{10}$$

$$N_1 = \sum_{j=1}^{w_i} (W_i^x(j)^2 + B_i^x(j)^2), \tag{11}$$

$$N_2 = \sum_{j=1}^{w_i} (W_{i-1}^x(j-s)^2 + B_{i-1}^x(j-s)^2). \tag{12}$$

The search range for s can be, for example, ± 64 pixels and for the estimation of x_i, the s value that maximizes Eq. (9) is used. Using row histograms, the value of y_i is estimated analogously, which results in the final translation vector of $\vec{v}_i = (x_i, y_i)$.

To incorporate temporal motion estimation Kalman filter is used [99]. To determine the weight of the current translation computation and the preceding trajectory, a novel heuristic is applied. The mean μ and standard deviation σ of the distances d between a pair of consecutive motion vectors (MVs) ($d = |\vec{v}_{i-1} - \vec{v}_i|$) were precomputed for manually registered images. Assuming Gaussian distribution and near zero mean, over 99% of the MVs are found to be within 3σ from the previous vector. A likely indicator of incorrect measurement is $d > 3\sigma$ and, in this case, the corresponding shift vector is discarded. On the contrary, if $d \leq 3\sigma$, the state of the Kalman filter is updated using $\vec{v}_i$ as the current state and d as the variance of the measurement. In both cases, the current state of the filter is used as the shift vector.

5.3 A Real-Time System for Capturing HDR Videos [79]

In a more recent work by Guthier et al., a unified real-time system for capturing, registering, stitching, and tone mapping HDR videos is presented [79]. Fig. 4 shows an overview of this HDR video processing system.

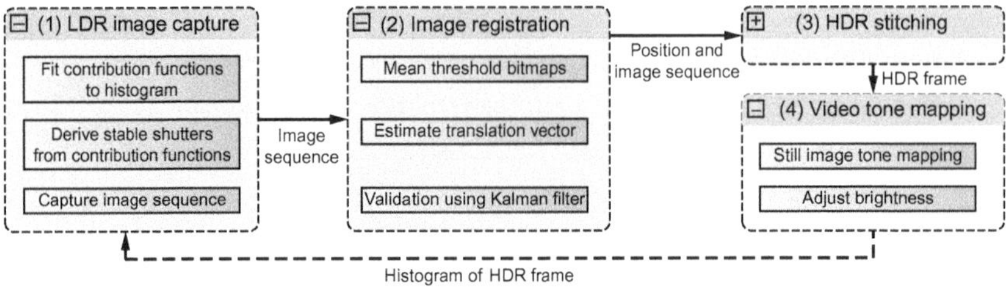

Fig. 4 An overview of the real-time HDR video processing system proposed by Guthier et al. [79]. *(Figure taken from the original work of the authors.)*

The aim of the first step is to only use the exposure times with the highest contribution of information to the generated HDR frame. By minimizing the number of captured images, the processing time is reduced, allowing higher frame rates. This is achieved by defining a contribution function $c_{\Delta t}(E)$ for each LDR image captured at exposure time Δt. The function computes the contribution of each image captured at Δt to the estimation of radiance E. It is defined as:

$$c_{\Delta t}(E) = w(f(E\Delta t)), \tag{13}$$

where E is the scene radiance, f is the camera response function, and w is a weighting function that measures the well-exposedness of pixel values.

The method computes the optimal exposure time sequence by using radiance histograms, which are generated as a by-product of tone mapping the previous frames. The details of using log-radiance histograms to compute optimal exposure times can be found in [101]. Briefly, the logarithm of the scene radiance histogram contains M bins, $(j = 1, \ldots, M)$, where each index j corresponds to the logarithm of a discrete radiance value: $b_j = \log(E_j)$. Each bin j contains $H(j)$ number of pixels in the HDR image having a log-radiance close to b_j. The bins have even spacing in the log domain, thus, $\Delta b = b_{j+1} - b_j$. Nonlogarithmic radiance values corresponding to two consecutive bins differ by a constant factor $\exp(\Delta b) = \exp(b_{j+1})/\exp(b_j) = E_{j+1}/E_j$. The optimal exposure times are selected such that the most frequent radiance values are well-exposed in at least one LDR image. This is obtained by ensuring that the peaks of the contribution functions in each exposure coincide with the histogram peaks as shown in Fig. 5. For a different exposure time $\Delta t'$, the corresponding contribution function is computed by simply shifting the original function to another position in the histogram. Therefore, the corresponding exposure times are computed by moving the contribution

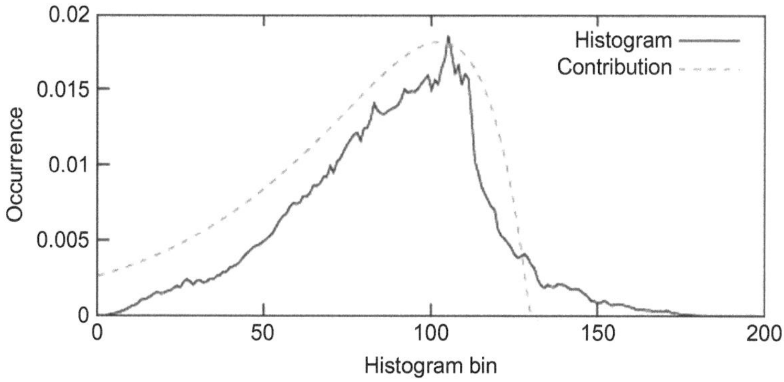

Fig. 5 The solid line represents an example of log-radiance histogram. The dashed line represents the contribution function in the log domain corresponding to the first shutter speed selected by the proposed method. The exposure is selected so that it captures the most frequently occurring radiance values best. *(Figure taken from B. Guthier, S. Kopf, W. Effelsberg, A real-time system for capturing HDR videos, in: Proceedings of the 20th ACM International Conference on Multimedia, ACM, 2012, pp. 1473–1476.)*

function over the histogram peaks and deriving the corresponding exposure time by using the following formula:

$$\Delta t' = \exp(\Delta b)^s \Delta t, \tag{14}$$

where the contribution vector is shifted by a number of s bins. The computation of new exposure times is repeated until the whole histogram is covered. With this approach, only those LDR images which accurately measure the scene radiance are captured.

The described algorithm for computing exposure times has two problems. The first problem is that the algorithm assumes perfect scene histograms. However, perfect histograms are not available for real-time videos. The problem of imperfect histograms is handled by choosing the first exposure time such that its contribution peak covers the highest radiance bin of the histogram. The reason for choosing the first exposure time in such a way is the fact that underexposed images have more accurate information than overexposed ones. This is because in underexposed images, dark pixels are a noisy estimate of the radiance in the scene, and this noise is unbiased. However, overexposed pixels always have a maximum pixel value, regardless of how bright the scene actually is.

The second problem is flicker that occurs due to the changes in exposure time sequence over time. Furthermore, if the camera is running in a sequence mode (i.e., a sequence of exposure parameters are sent only once and are then repeatedly used throughout the capture), any change in exposure time sequence requires an expensive retransmission of the parameters to the camera. To avoid this, a stability criterion is enforced. This criterion ensures that the new exposure time sequence is only retransmitted when the computed sequence is different from the previous sequence for a number of successive frames. With this approach, temporal stability is also achieved.

The second stage of the proposed system is the alignment of the successive frames, which utilizes the algorithm described in Section 5.2. Once the frames are aligned, HDR stitching is performed using the radiance map recovery technique of Debevec and Malik [4].

The last stage of the system is tone mapping of generated HDR frames. The aim is to apply a tone mapping operator designed for still images while taking measures to avoid flicker. To this end, the average brightness of tone mapped frames is adjusted to be closer to the preceding frames.

6. LOCAL VIDEO DEGHOSTING

Unlike global methods, local deghosting algorithms consider both camera and object motion. Therefore, in general, the algorithms presented in this section are more sophisticated than the former.

6.1 High Dynamic Range Video [80]

This algorithm is considered to be the first approach to generate HDR videos using off-the-shelf cameras by alternating frames between short and long exposures. It is comprised of the following three steps: (1) frame capture using automatic exposure control, (2) HDR stitching across neighboring frames, and (3) temporal tone mapping for display.

6.1.1 Capture

In the capture stage, the exposure settings alternate between two values that are continuously updated to incorporate scene changes. This is done by calculating scene statistics on a subsampled frame. The ratio between the exposures varies from 1 to a maximum specified by the user. It has been observed that the maximum value of 16 generates good results, in terms of both recovering the dynamic range and compensating for the motion.

6.1.2 HDR Stitching

The goal of this stage is to create an HDR frame by using information from the neighboring frames. First, gradient-based optical flow is used to *unidirectionally* warp the previous/next frames to the current frame. Then the warped frames are merged with the current frame in the well-exposed regions of the latter. The over- and underexposed regions of the current frame are *bidirectionally* interpolated using optical flow followed by a hierarchical homography algorithm to improve the deghosting process. The details of this workflow are explained as follows.

Unidirectional Warping. In this stage, a flow field is computed between the current frame, L, and its neighboring frames S^- and S^+. However, to compensate for the exposure differences, the neighboring frames are first "reexposed" using the camera response curve and the exposure value of the current frame. For computing the optical flow, a variant of the Lucas and Kanade technique [102] in a Laplacian pyramid framework [103] is used. After computing the optical flow, the neighboring frames are warped to obtain S_U^- and S_U^+.

Bidirectional Warping. This stage is used to estimate the overexposed regions of the current frame (or underexposed if the current frame is a short frame). Because the current frame cannot inform the flow computation in these regions, a bidirectional flow field is computed directly between the previous and the next frames. Using these flow fields, an intermediate frame is computed and this is called as $\tilde{S}_B$. This intermediate frame is then reexposed to match its exposure to the current frame. This reexposed frame is called $\tilde{L}_B$. Next, an hierarchical homography is computed between the current frame L and $\tilde{L}_B$. The result of this homography also results in a flow field due to its hierarchical nature. This flow field is added to the original bidirectional flow field to compute the final flow field. S^- and S^+ are then warped using this flow field to obtain S_B^- and S_B^+.

Radiance Map Recovery. The next step is to combine the current frame with the four auxiliary frames that were generated. To this end, the following steps are applied, assuming the current frame is a long exposure:

- Using the response function, for all images (L, S_U^-, S_U^+, S_B^-, and S_B^+) compute radiance images ($\hat{L}$, $\hat{S}_U^-$, $\hat{S}_U^+$, $\hat{S}_B^-$, and $\hat{S}_B^+$).
- Overexposed pixels in the final radiance map are filled in with bidirectionally interpolated pixels from $\hat{S}_B^-$ to $\hat{S}_B^+$. To avoid possible blurring, pixels from solely the previous frame $\hat{S}_B^-$ are used. Only if they happen to be inconsistent with the current frame (too low to saturate in the current frame) then the pixels from $\hat{S}_B^+$ are used.

- The radiance map in other regions is computed as a weighted blend:

$$R = \frac{f_{WM}(p, p^-)p^- + f_{WM}(p, p^+)p^+ + f_W(p)p}{f_{WM}(p, p^-) + f_{WM}(p, p^+) + f_W(p)},$$

  where the pixels p, p^-, and p^+ come from $\hat{L}$, $\hat{S}_U^-$, and $\hat{S}_U^+$, respectively. The weighting function $f_W(.)$ is based on Mitsunaga and Nayar [104] and $f_M(.)$ is used to attenuate large pixel value differences. It is defined as:

$$f_M(\sigma) = \begin{cases} 2\left(\frac{\sigma}{\sigma_{\max}}\right)^3 - 3\left(\frac{\sigma}{\sigma_{\max}}\right)^2 + 1 & \text{if } \sigma < \sigma_{\max}, \\ 0 & \text{otherwise.} \end{cases}$$

  Here, $\sigma_{\max}$ is set to a value that corresponds to 16 gray levels in the longest exposure. Finally, $f_{WM}(p, q) = f_M(|p - q|)f_W(p)$.

  If current exposure is the short one, the same algorithm is used but in the second step underexposed pixels are discarded.

6.1.3 Temporal Tone Mapping

For tone mapping the generated video, Reinhard et al.'s photographic tone mapping operator is used [105]. The computation of log-average luminance is slightly modified to include multiple frames in the spirit of reducing temporal flicker:

$$\bar{L}_w = \exp\left(\frac{1}{N}\sum_{x,y,i}\log(\epsilon + L_w(x, y, i))\right), \tag{15}$$

where (x, y) are the spatial indices, i is the frame index, and $L_w(x, y, i)$ denotes the radiance value at pixel (x, y) in frame i. Only the current frame and the previous frame are used during this spatiotemporal averaging. Finally, ϵ is a small value used to avoid singularity for zero radiance pixels.

6.2 Video Matching [82]

Sand and Teller introduced an algorithm for spatiotemporal alignment of a pair of videos [82]. The input of the proposed algorithm is two videos, which may have been recorded with different exposures. Of these two videos, one is marked as the primary (reference) and the other as the secondary. The output is a modified secondary video whose pixels have been brought to temporal and spatial alignment with the primary video.

The proposed method is not specialized to HDR video, but it has been shown that it can be used to create HDR videos as well.

6.2.1 Spatial Alignment

Given a primary input frame I_1 and a secondary input frame I_2, the image alignment algorithm begins by detecting the feature points in both frames using Harris corner detector [106]. The initial matches from I_1 for each feature point in I_2 are found according to the similarity of nearby pixel values around each feature point.

In the next step, the best matching correspondences are found by maximizing a weighting function, w_i, that includes two terms namely pixel matching probability, P_i, and motion consistency probability, M_i:

$$w_i = P_i M_i. \tag{16}$$

The pixel matching probability is measured by calculating a pixel matching score across a square region R around the corresponding feature points in two frames. However, instead of comparing the primary frame with the secondary one directly, two auxiliary frames namely, I_2^{min} and I_2^{max}, are computed by taking the minimum and maximum of 3×3 regions around each pixel. This makes the algorithm more robust against small differences between pixel values:

$$\sum_{(x,y)\in R} \max\left(0, I_1(x,y) - I_2^{\mathrm{max}}(x+u, y+v), I_2^{\mathrm{min}}(x+u, y+v) - I_1(x,y)\right), \tag{17}$$

where (u, v) is the offset between the points, which was computed earlier. This dissimilarity score is aggregated over each color channel to determine the final dissimilarity score of each correspondence, d_i. The final pixel matching probability is computed as:

$$P_i = \mathcal{N}(d_i, \sigma_{\mathrm{pixel}}^2), \tag{18}$$

where $\mathcal{N}$ is a zero-mean normal distribution and σ_{pixel} is set to 2.

The motion consistency probability term measures the smoothness of the offset vectors. To this end, the authors first estimate a dense motion field, $u(x, y)$ and $v(x, y)$, from the MVs between the correspondences. This dense field is computed by a modification of the locally weighted linear regression algorithm [107] and is called adaptive locally weighted regression. The MV at each pixel is computed by fitting a Gaussian kernel to the nearby feature points. The width of this kernel is inversely proportional to the number of nearby feature points. That is, if a pixel is surrounded by many feature points, the kernel width is reduced. On the other hand, if the feature points around a pixel are sparse, larger kernel widths are employed. The MV is then computed as a weighted average of these kernel functions with the weight of each kernel determined by the correspondence score, w_i. Motion consistency is then given by the similarity of these estimated MVs, $(\hat{u}_i, \hat{v}_i)$, from the previously assigned MVs, (u_i, v_i):

$$M_i = \mathcal{N}\left(\sqrt{(u_i - \hat{u}_i)^2 + (v_i - \hat{v}_i)^2}, \sigma^2_{\text{motion}}\right). \tag{19}$$

After the initial matches are found, the correspondences between the feature points are improved iteratively by checking the pixel matching score of the pixel at location predicted by $(\hat{u}_i, \hat{v}_i)$, and several feature points around which are detected by the corner detector. For each candidate correspondence, the proposed algorithm applies a local motion optimization using the KLT method [102, 108]. When the iteration converges to a good set of correspondences, a dense correspondence field is obtained using the adaptive locally weighted regression.

6.2.2 Temporal Alignment

The alignment algorithm discussed earlier is used as the core of a video matching algorithm, which tries to align two videos temporally and spatially. First, each frame in the secondary video is matched to a nearby frame in the primary video. This is accomplished by minimizing the following cost:

$$D_{i,j} = \lambda p^2_{i,j} + m^2_{i,j}, \tag{20}$$

where i and j are the frame indices, $p_{i,j}$ is a parallax measure, and $m_{i,j}$ is simply the mean correspondence vector magnitude. λ, which is set to five, is used to control the relative influence of the two terms.

The parallax measure is a measure of depth discontinuity between the frames. Given a pair of correspondences, the difference of the Euclidean distance between them in the first image and the second image is proposed as a suitable measure of these discontinuities. $p_{i,j}$ is computed by averaging this difference for all pairs of correspondences.

Finally, instead of comparing a given frame in the secondary video with all frames in the primary video, an optimized search is employed. This involves computing $D_{i,j}$ for several nearby frames and then fitting a quadratic regression function to the obtained values. The search is then conducted around the frames for which this quadratic function is minimum. Once the matching frame is found, a dense flow field is computed by using the image alignment algorithm described earlier.

6.2.3 *HDR Reconstruction*

In order to produce an HDR video, this algorithm is applied on a pair of input videos with different exposures after local contrast and brightness normalization [109]. The output pair of aligned videos may be merged using a standard HDR reconstruction method, such as the algorithm of Debevec and Malik [4]. In order to obtain a satisfactory result, the viewpoint must not change significantly between the two input videos. In addition, this algorithm is more successful at recovering the dynamic range of the background rather than the moving objects because the dense correspondence fields are smoothed out to mostly account for the global motion in the input frames.

6.3 HDR Video Through Fusion of Exposure Controlled Frames [110]

Youm et al. [110] proposed a technique that employs a novel exposure control method in addition to ghost removal. The proposed method adaptively selects the exposure times of the individual frames during the acquisition of an LDR video with alternating long and short exposures. After the acquisition, the pyramid-based image fusion method of Burt and Adelson [111] is used for the fusion of the captured frames.

The exposure control method adaptively selects the exposure time of the next frame, using the information available in the current frame, and it is based on the adaptive dynamic range imaging method of Nayar and Branzoi [112]. The algorithm uses two user-set parameters: the saturation level Z^{sat} and the desired intensity level range $[Z^{\text{des}} - \Delta Z, Z^{\text{des}} + \Delta Z]$. The exposure time of the current frame, Δt_i, is determined from the exposure time of the previous frame with the similar exposure, Δt_{i-2}, using the following set of rules assuming that the current frame is a short exposure:

$$\Delta t_i = \begin{cases} 0.5\Delta t_{i-2}, & \text{if } M^{\text{avg}} \geqslant Z^{\text{sat}}, \\ \Delta t_{i-2}\frac{Z^{\text{des}}}{M^{\text{avg}}}, & \text{if } Z^{\text{sat}} > M^{\text{avg}} \geqslant Z^{\text{des}} + \Delta Z, \\ \Delta t_{i-2}, & \text{if } Z^{\text{des}} + \Delta Z > M^{\text{avg}} \geqslant Z^{\text{des}} - \Delta Z, \\ \Delta t_{i-2}\left(0.5\frac{Z^{\text{des}}}{M^{\text{avg}}} + 0.5\right), & \text{if } Z^{\text{des}} - \Delta Z > M^{\text{avg}}, \end{cases} \tag{21}$$

where M^{avg} is the average intensity in the bright regions of the scene for short-exposure frames. For long exposures, an analogous set of rules is applied.

In the exposure fusion step, the contributions of a pair of long- and short-exposure frames to the output frame depend on the well-exposedness mask M^{we} and the ghost mask M^{g}. Given a pixel value $Z_i(x, y)$ from the long exposure and $Z_{i-1}(x, y)$ from the short exposure, the well-exposedness mask is defined as:

$$M_i^{we} = \begin{cases} 1, & \text{if } Z_i(x,y) < Z^{cut}, \\ \frac{255-Z_i(x,y)}{255-Z^{cut}}, & \text{if } Z_i(x,y) \geqslant Z^{cut}, \end{cases} \tag{22}$$

where Z^{cut} is the cut-off intensity for the well-exposedness map. The ghost mask M^{g} is calculated by comparing the current exposure with the previous corresponding exposure. It takes the value of zero if the absolute pixel intensity difference is larger than 20 for 8-bit input frames and one otherwise.

The output frame is a weighted pyramid-based blending [111] of short and long exposures. The blending weight of the long exposure is $M_i^{we} \cdot M_i^{g}$ and the blending weight of the short exposure is $1 - M_i^{we} \cdot M_i^{g}$.

6.4 High Dynamic Range Video With Ghost Removal [86]

Recognizing that dense optical-flow based HDR video [80] cannot cope well with occlusions and fast-moving objects, Mangiat and Gibson proposed a novel method that uses block-based motion estimation instead [86]. Their method is comprised of four stages: (1) block-based motion estimation, (2) MV refinement in saturated blocks, (3) HDR reconstruction, and (4) artifact removal. The details of each stage is explained as follows.

6.4.1 Block-Based Motion Estimation

The block-based motion estimation involves two frames, with one being a *reference* frame and the other the *current* frame. The goal is to find for each block in the current frame the best matching block in the reference frame. To this end, the current frame is discretized into square blocks of size 16×16 and a sliding window of equal size is traversed over the reference frame. The traversal stops when, according a predefined error metric, the difference between the two blocks is smaller than a threshold. Note that this process is commonly used during video compression to exploit the temporal coherence between frames. The difference in the HDR setting is that the consecutive frames are differently exposed, which makes finding correspondences more difficult.

To address the exposure difference, Mangiat and Gibson first reexpose the short exposure to match the long exposure (or vice versa):

$$Z_l = g^{-1}(g(Z_s) - \ln \Delta t_s + \ln \Delta t_l), \tag{23}$$

where g is the logarithm of the inverse camera response function, i.e., $g(x) = \ln f^{-1}(x)$. Next, two unidirectional motion fields are computed between the current frame (long exposure) and the neighboring frames (previous and next short exposures). To this end, the authors use the H.264 JM Reference Software[1] with Enhanced Predictive Zonal Search (EPZS) [113]. This algorithm optimizes finding correspondences by first considering the most likely predictors for the MVs, such as the median MVs of neighboring blocks and MVs of the neighboring blocks in the previous frame. The algorithm stops if these predictors yield an error term that is smaller than an adaptively set threshold. Otherwise, it continues searching using a fixed pattern. The error term is defined as the sum of absolute differences (SAD):

$$\mathrm{SAD}_{I_1,I_2}(x_1, y_1, x_2, y_2) = \sum_{m=0}^{N-1} \sum_{n=0}^{N-1} |I_1(x_1 + m, y_1 + n) - I_2(x_2 + m, y_2 + n)|. \tag{24}$$

Here, I_1 and I_2 represent the images, N the square block size, and (x_1, y_1) and (x_2, y_2) are the coordinates of the blocks to be compared. Typically, the coordinate pairs that yield the smallest SAD value are assumed to be matching and an offset, (v_x, v_y), is computed from their difference. This difference is termed as the motion vector. For color images, this search is conducted for both luma and chroma components. Mangiat and Gibson perform this search for both the previous and the next frames of the current frame, and for each block, choose the frame that has the smallest SAD value. To this end, two pieces of information are stored for each block of the current frame: (1) the MV and (2) the label of the previous or the next frame for which this MV is valid. It is also stated by the authors that the estimated MVs can be used to estimate the global camera motion (if any) by using a least squares approach.

EPZS algorithm works well for properly exposed regions. However, for saturated blocks, the current frame does not have sufficient information to guide this unidirectional search. A saturated block is defined as one

[1] http://iphome.hhi.de/suehring/tml.

for which more than 50% of its pixels are above (or below) a threshold. For these blocks, a bidirectional search is performed directly between the previous and the next frames. However, the MVs of the neighboring *unsaturated* blocks of the current frame inform this search to ensure that the MV of a saturated block is not vastly different from the MVs of neighboring unsaturated blocks. Assuming that the block (x, y) is saturated in the current frame,[2] the coordinates of the matching blocks in the previous and the next frame are computed by $(x_1, y_1) = (x, y)+(v_x, v_y)$ and $(x_2, y_2) = (x, y)-(v_x, v_y)$, respectively. These coordinates can be computed by finding an (v_x, v_y) value which minimizes:

$$\text{Cost} = \text{SAD}_{I_p,I_n}(x_1, y_1, x_2, y_2) + \lambda|(v_{mx}, v_{my}) \pm (v_x, v_y)|, \qquad (25)$$

where I_p and I_n denote the previous and the next frames, (v_{mx}, v_{my}) denote the *median* of the MVs of neighboring unsaturated blocks, and λ is a factor, which determines the relative contribution of the first (data) and the second (smoothness) terms. To compute the median MV, the labels of unsaturated blocks in the 5×5 neighborhood of the saturated block are tallied, and the frame (previous or next) with the highest count is selected as the reference. The median MV is then found as the median of the MVs of blocks that are labeled with this reference. Finally, $\pm$ represents the operation to be applied depending on whether the reference is selected as the previous or the next frame. Once a saturated block is fixed, it is marked as unsaturated so that it can also inform neighboring saturated blocks. This way larger holes (saturated regions) are gradually filled.

6.4.2 MV Refinement in Saturated Blocks

It is possible that the previous matching algorithm produces incorrect correspondences between blocks. This problem is most pronounced for saturated blocks as for these blocks, the information from the current frame is not used at all. However, although these saturated blocks are not informative, one can still estimate the minimum (or maximum) pixel values that could exist in a matching block. Assuming that the current image is the long exposure and the reference is either the previous or the next short exposure, a saturated pixel of the current exposure should not have been matched with a reference pixel that is too dark to saturate in the current exposure. If this happens, this match is likely to be erroneous.

[2] (x, y) denotes the pixel coordinates of the top-left corner of the block.

Therefore, in this stage of the algorithm, the authors determine the minimum pixel value that is valid as:

$$Z_s^* = g^{-1}(g(Z_{\max}) - \ln \Delta t_l + \ln \Delta t_s). \tag{26}$$

Any pixel that is smaller than this value in the reference exposure cannot be the correct match for the saturated pixel in the current exposure. Similarly, if the current exposure is a short exposure, a maximum valid value is computed as:

$$Z_l^* = g^{-1}(g(Z_{\min}) - \ln \Delta t_s + \ln \Delta t_l). \tag{27}$$

The erroneous pixels are identified to be those that are smaller than Z_s^* if the current frame is a long exposure and greater than Z_l^* if it is a short exposure. Once these pixels are located, their MVs are updated using their eight immediate neighbors and eight nearest neighbors from the adjacent blocks. This amounts to choosing an MV that maximizes the following weight function:

$$w(p, q_k) = r(p, q_k)c(p, q_k) = \frac{1}{\|p - q_k\|} \frac{1}{|I_r(p + MV_k) - I_r(q_k + MV_k)|}, \tag{28}$$

where I_r represents the reference exposure and q_k the 16 candidate pixels from which one of the MVs will be selected. The $r(p, q_k)$ term represents the spatial distance and therefore, nearby pixels are favored. The $c(p, q_k)$ term, on the other hand, is a color similarity term based on the idea that pixels that are similar in color should have similar MVs. The refinement process is illustrated by an example in Fig. 6.

Fig. 6 The long (current) exposure shown on the *left* is combined with the estimated exposure using MVs of adjacent short exposures. This results in some artifacts near the boundaries of the vehicle. These artifacts are removed after the MV refinement process. *(Figure taken from S. Mangiat, J. Gibson, High dynamic range video with ghost removal, in: SPIE Optical Engineering+ Applications, International Society for Optics and Photonics, 2010.)*

6.4.3 HDR Reconstruction

This stage of the algorithm merges the current exposure and the estimated exposure using Debevec and Malik's algorithm [4]:

$$\ln E(x, y) = \frac{\sum_{i=1}^{P} w(Z_i(x, y))(g(Z_i(x, y)) - \ln \Delta t_i)}{\sum_{i=1}^{P} w(Z_i(x, y))}, \tag{29}$$

where w is a triangular weighting function and P is the number of exposures, respectively. The generated HDR image is then tone mapped using Reinhard et al.'s photographic tone mapping operator [105].

6.4.4 Artifact Removal

Given the block-based nature of the algorithm, it is possible that blocking artifacts will appear at the block boundaries. In this stage, the color information of the current frame is used to fix these artifacts. This is achieved by filtering the tone mapped HDR image, using a cross-bilateral filter (CBF) in which the edge information is taken from the current frame [114]. This produces a filtered image with edges smoothed out if they do not exist in the current frame, such as the edges caused at block boundaries. To identify these regions, the filtered image and the original tone mapped HDR image are compared using perceptual color difference and SSIM [115] metrics. If the color difference is greater than a threshold or if the SSIM score is less than a threshold, the tone mapped HDR pixel is replaced by the corresponding pixel in the filtered image. However, if the current frame is saturated for these pixels, the original tone mapped pixel value is not changed.

6.5 Spatially Adaptive Filtering for Registration Artifact Removal in HDR Video [88]

A follow-up work by Mangiat and Gibson proposes to improve the artifact removal stage of their previous algorithm. As discussed in the previous section, the artifact removal stage involved applying a CBF to fix the block artifacts:

$$\mathrm{CBF}[J]_p = \frac{1}{W_p} \sum_{q \in \mathcal{S}} G_{\sigma_s}(\|p - q\|) G_{\sigma_r}(|I_p - I_q|) J_q, \tag{30}$$

where $\mathcal{S}$ is a neighborhood around pixel p. Here, the tone mapped image J is filtered by using the edge information in frame I. G_{σ_s} denotes a Gaussian function with standard deviation σ_s and likewise for G_{σ_r}. W_p represent the sum of all weights. This process blurs across the edges of image J if those edges are not present in image I, e.g., the edges caused by blocking artifacts.

Recognizing that artifacts are most likely to occur in large motion regions, the authors propose the modify the standard deviation of the range kernel as:

$$\sigma_r(p) = \alpha \max_{q \in \mathcal{S}}(\|MV_q\|), \tag{31}$$

where α is a user parameter used to adjust the amount of smoothing.

However, this modification cannot fix registration errors that may occur in saturated regions as, in these regions, the current frame does not have any edge information. Therefore, the authors propose a second modification to use the edge information in the tone mapped HDR frame directly for the saturated regions of the current frame. This is formulated as follows:

$$\mathrm{HDR}[J, I]_p = \begin{cases} \frac{1}{W_p} \sum_{q \in \mathcal{S}} G_{\sigma_s}(\|p - q\|) G_{\sigma_I(p)}(|I_p - I_q|) J_q, & \text{if } 10 < I_p < 230, \\ \frac{1}{W_p} \sum_{q \in \mathcal{S}} G_{\sigma_s}(\|p - q\|) G_{\sigma_I(p)}(|I_p - I_q|) G_{\sigma_J(p)}(|J_p - J_q|) J_q, & \text{otherwise.} \end{cases} \tag{32}$$

In other words, in regions where the current frame is well-exposed, the tone mapped frame has no influence on the filter. In these regions, the adaptive version of the CBF is used. However, in the saturated regions of the current frame, the tone mapped frame itself is used to smooth out potential registration errors. An additional benefit of this approach is that it obviates the need to perform perceptual color similarity or SSIM computations as the final tone mapped frame becomes the filtered frame itself.

6.6 Filter-Based Deghosting for Exposure Fusion Video [83]

In their work, Chapiro et al. [83] present a method to deal with ghost artifacts based on the exposure fusion approach of Mertens et al. [116]. The proposed method addresses small-scale movements rather than fast object and camera motions. For improving the quality of the input video by eliminating ghost artifacts on object boundaries, a ghosting term is included in addition to the contrast, saturation, and well-exposedness terms of the exposure fusion algorithm. This term reduces the contribution of the motion-affected pixels.

The proposed method begins with applying a multiresolution global registration to each pair of consecutive input frames. In order to detect the pixels that may cause ghost artifacts, low-pass Gaussian filters are applied to the input frames, followed by high-pass Laplacian filters. The ghost term for a specific pixel is calculated by subtracting the SAD from one around a rectangular region in each pair of input frames.

We note that this method does not generate an HDR video, but rather, a better exposed LDR video by directly merging the input frames similar to Youm et al. [110]. Therefore, it does not require information, such as the camera response and exposure times.

6.7 Toward Mobile HDR Video [81]

Castro et al. [81] presented a method of reconstruction of HDR video for hand-held cameras as an extension of a similar earlier work [84]. Because the method is based on histograms, it is of low computational cost and thus, suitable for less powerful processors, such as mobile phones. The proposed algorithm initially captures a multiexposure sequence on Nokia N900. Then the captured exposures are transferred to a desktop computer to generate HDR videos. Finally, the constructed HDR videos are tone mapped to enable visualization on LDR displays. The proposed method of generating HDR video has three steps: (1) photometric calibration, (2) multiresolution alignment, and (3) radiance map estimation.

For photometric calibration, the method takes a sequence of three (or possibly more) exposures F^i as input, where $F^i = \{F_1^i, F_2^i, F_3^i\}$. Exposure F_1^i is constant for all i and is determined before capture, using the auto-exposure feature of the camera. F_2^i and F_3^i are exposures of twice and half the exposure value of F_1^i, respectively, for all i. Photometric calibration relies on pixel correspondences between pixels from different frames. The authors assume that exposure changes preserve monotonicity of pixel values and define the radiance mapping $M_{P,Q}$ between two consecutive frames, P and Q as:

$$M_{P,Q}(p_i) = q_i, \forall i, \tag{33}$$

where p_i and q_i represent the corresponding pixels between two frames. Radiance map is then reconstructed by using the approach described in Mitsunaga and Nayar [104].

However, prior to constructing the radiance map, the exposures must be first brought into global alignment. This global alignment is performed by applying Ward's MTB registration method [10].

Finally, the removal of ghost artifacts is achieved by analyzing the variance of radiance values over the corresponding pixels of aligned images in F^i. Four radiance maps, $\hat{F}_j^i$ for $j = 1, 2, 3, 4$ are generated for each F^i. The first three radiance maps are simply the original frames converted to the radiance domain, and the last one is the reconstructed HDR image after global alignment. The first three radiance maps are then updated based on the magnitude of the variance. If the pixel variance is low, the HDR frame is given more weight, whereas for high-variance pixels, the original pixel values are prioritized.

6.8 Patch-Based High Dynamic Range Video [85]

The main contribution of this method, which arguably produces the highest quality deghosting results, is a judicious combination of optical-flow [80] and patch-based [62] HDR image synthesis algorithms. For the capture process, the input is assumed to be N LDR images alternating between M different exposures $(L_{\text{ref},1}, L_{\text{ref},2}, \ldots, L_{\text{ref},N})$. The task is to generate the N HDR frames $(H_n, n \in 1, \ldots N)$. For each frame, additional $(M-1)$ missing frames are synthesized (for all but the reference frame).

Although [62] works very well for still images, it is not suitable for HDR video due to lack of temporal coherency. Also, this algorithm fails if a large region of the image is under- or over-exposed. Thus, the direct application of [62] is not suitable. Therefore, the following modified energy function is proposed:

$$\begin{aligned} E(\text{all } L_{m,n}) = \sum_{n=1}^{N} \sum_{p \in \text{pixels}} & [(\alpha_{\text{ref},n_{(p)}} \cdot (h(L_{\text{ref},n})_{(p)} - H_{n(p)})^2 \\ & + (1-\alpha_{\text{ref},n_{(p)}}) \cdot \sum_{m=1, m\neq \text{ref}}^{M} \Lambda(L_{m,n})(h(L_{m,n})_{(p)} - H_{n(p)})^2 \\ & + (1-\alpha_{\text{ref},n_{(p)}}) \cdot \sum_{m=1}^{M} \text{TBDS}(L_{m,n}, L_{m,n-1}, L_{m,n+1}))], \end{aligned} \quad (34)$$

where $h(.)$ converts LDR images to the linear radiance domain, α is an approximation of how well a pixel is exposed, and H_n is an HDR frame. Λ in the second term is weighted triangle function used for merging [4]. The third term, temporal bidirectional similarity (TBDS), adds temporal coherence between the estimated exposures:

$$\text{TBDS}(L_{m,n}, L_{m,n-1}, L_{m,n+1}) = \text{BDS}(L_{m,n}, L_{m,n-1}) + \text{BDS}(L_{m,n}, L_{m,n+1}), \quad (35)$$

with BDS is proposed in [117]:

$$\text{BDS}(T,S) = \frac{1}{|S|} \sum_{p \in \text{pixels}} \min_{i \subset f_S^T(p) \pm w_S^T(p)} D(s(p), t(i)) + \frac{1}{|T|} \sum_{p \in \text{pixels}} \min_{i \subset f_T^S(p) \pm w_T^S(p)} D(t(p), s(i)),$$

where patch center in pixel p is marked by $s(p)$ (source) and $t(p)$ (target) and function D is the sum of the squared differences (SSD) between s and t. Here, $f_S^T(p)$ approximates the optical flow at pixel p from the source to target and $w_S^T(p)$ scales the search window. Note that for improved efficiency, the search is conducted around the regions indicated by an initial optical-flow estimation. In summary, the proposed method contains the following three main steps:

Initial motion estimation. During this stage, motion is roughly estimated in both directions for consecutive frames ($f_S^T(p)$ and $f_T^S(p)$). For global motion estimation, a planar similarity model and for local motion, optical-flow is used. The accuracy of this step influences the size of the search window in the next step.

Search window map computation. A window size of $w_S^T(p)$ and $w_T^S(p)$ is computed for every flow vector. This window is used to refine the initial motion estimate.

HDR video reconstruction. A two-stage iterative method is utilized in which the first stage minimizes the last term in Eq. (34), and the second stage minimizes the first two terms. These two stages are performed in succession until convergence. This produces all of the missing frames and the corresponding HDR image at each time instance.

6.9 A MAP Estimation Framework for HDR Video Synthesis [87]

One of the main difficulties in HDR video synthesis is estimation of accurate MVs between a sequence of multiexposures captured at different shutter speeds. Li et al. [87] proposed a maximum a posteriori (MAP) estimation framework for HDR video synthesis that does not use optical flow calculation for motion estimation. The proposed algorithm is an extension of the rank minimization-based HDR imaging (RM-HDR) [42, 50, 118].

The framework is based on an image acquisition model defined as follows: It is assumed that a pixel value z is defined as $z = g(a\Delta t) + n$, where a is the radiance value, Δt is the exposure time, $g(\cdot)$ is the response function, and n is a Gaussian random noise. The conditional probability of z given a is defined as:

$$\begin{aligned} f(z|a) &= \frac{d}{dz} Pr\{g(a) + n < z\} \\ &= \frac{d}{dz}\phi\left(\frac{z - g(a)}{\sigma_n}\right) \alpha \exp\left(-\frac{(z - g(a))^2}{2\sigma_n^2}\right), \end{aligned} \tag{36}$$

where σ_n is the standard deviation of n, and ϕ is the cumulative distribution function of the standard Gaussian distribution.

The model generates an aligned background and computes the foreground radiance simultaneously. To solve the problem of scene motion, each exposure is separated into foreground and background regions. Initially, a binary matrix $\mathbf{S} \in \{0, 1\}^{KxN}$, where K is the number of pixels in each frame and N is the number of exposures is constructed as:

$$S_{i,j} = \begin{cases} 1, & \text{if pixel } (i,j) \text{ is in the foreground,} \\ 0, & \text{if pixel } (i,j) \text{ is in the background,} \end{cases} \tag{37}$$

where (i,j) corresponds to the ith pixel in the jth frame.

A probabilistic model for $\mathbf{S}$ is then constructed as a Markov random field under the assumption that interactions of a given pixel with all others are entirely described by its neighbors. The probability distribution function that can be considered as the prior probability of $\mathbf{S}$ is then defined as:

$$f(\mathbf{S}) \; \alpha \; \exp\left(\sum_{(i,j)\in\mathcal{E}} \lambda_{ij} l_i l_j + \sum_{i\in\mathcal{V}} \lambda_i l_i \right), \tag{38}$$

where $\mathcal{V}$ and $\mathcal{E}$ are the set of vertices and edges of the graphical representation of $\mathbf{S}$, respectively, λ_{ij} and λ_i are edge interaction costs and vertex weights, and $l_i \in \{-1, 1\}$ is the latent variable corresponding to the ith pixel in $\mathbf{S}$. We refer the reader to Lee et al. [119] for a more detailed description of $\mathbf{S}$ and setting the values of parameters λ_{ij} and λ_i.

Next, the radiance matrix $\mathbf{A}$ for the synthesized HDR video is defined as a combination of the foreground $\mathbf{C}$ and background $\mathbf{B}$ matrices:

$$\mathbf{A} = \mathcal{P}_{\mathbf{s}}(\mathbf{C}) + \mathcal{P}_{\mathbf{s}^c}(\mathbf{B}), \tag{39}$$

where $\mathcal{P}_{\mathbf{x}}(\mathbf{Y})$ is a sampling operator defined as:

$$[\mathcal{P}_{\mathbf{x}}(\mathbf{Y})]_{ij} = \begin{cases} Y_{ij}, & \text{if } X_{ij} = 1, \\ 0, & \text{otherwise.} \end{cases} \tag{40}$$

Since $\mathbf{B}, \mathbf{C}$, and $\mathbf{S}$ are unknown, they need to be estimated. Their estimation is defined as a MAP estimation problem:

$$(\hat{\mathbf{S}}, \hat{\mathbf{B}}, \hat{\mathbf{C}}) = \underset{\mathbf{S},\mathbf{B},\mathbf{C}}{\operatorname{argmax}} f(\mathbf{S}, \mathbf{B}, \mathbf{C}|\mathbf{D}), \tag{41}$$

where $D := [vec(I_1), vec(I_2), \ldots, vec(I_N)]$ corresponds to the scene observation matrix, where each I_k denotes the observed radiance map of the kth frame obtained by using the inverse camera response function as defined in Debevec and Malik [4], and $f(\mathbf{S}, \mathbf{B}, \mathbf{C}|\mathbf{D})$ corresponds to the joint probability of **S**, **B** and **C**, given **D**. The authors rewrite the estimation problem defined in Eq. (41) as:

$$(\hat{\mathbf{S}}, \hat{\mathbf{B}}, \hat{\mathbf{C}}) = \underset{\mathbf{S},\mathbf{B},\mathbf{C}}{\operatorname{argmin}} \frac{1}{2} \|\mathcal{P}_{\Omega}(\mathbf{D} - \mathbf{B})\|_F^2 + \alpha \|\mathbf{B}\|_{\star} + \beta \|\mathbf{S}\|_1 + \gamma \|\mathbf{W}\text{vec}(\mathbf{S})\|_1 + \|\mathbf{V}\text{vec}(\mathcal{P}_S(\mathbf{C} - \mathbf{E}))\|^2, \tag{42}$$

where Ω is the support of the well-exposed background, **W** is the weighting matrix responsible for interactions between neighboring pixels, and $\alpha, \beta, \gamma > 0$ are constant parameters used to control the relative importance between each term. To solve Eq. (42), an iterative optimization over each variable, **S**, **B**, and **C**, is performed until convergence. More detailed description on the derivation and solution of Eq. (42) can be found in [87].

6.10 Motion Aware Exposure Bracketing for HDR Video [120]

Gryaditskaya et al. [120] introduced a real-time adaptive selection of exposure settings and an off-line algorithm for the reconstruction of HDR video, which is specifically tailored for processing the output of the proposed adaptive exposure approach.

First, the authors present the results of a psychophysical experiment for comparing the relative importance of limited dynamic range vs. ghosting artifacts, because there is a trade-off between the two in multiexposure HDR acquisition. As a result of this experiment, the authors find that the effect of ghosting artifacts is much more significant than that of the limited dynamic range on the perceived quality.

Next, the authors introduce their metering algorithm for the detection of optimal exposure times for minimizing ghosting, while maximizing dynamic range. The first two frames with long and short exposures are captured with the built-in metering system of the camera. The pixel intensities in the frame with shorter exposure are adjusted to the longer exposure. Then, the Hierarchical Diamond Search of Urban et al. [121] is used to estimate the motion between frames.

After calculating the motion field, the dominant MV for a frame is found by using the block-based recursive weighted least-squares method. The blocks which follow the dominant MV are marked as dominant motion regions (DMRs), while other blocks are marked as local motion regions (LMRs).

Placing priority on deghosting while extending the dynamic range, a requirement found from the psychophysical experiment, is satisfied by fulfilling the following two conditions while adaptively selecting the best estimate $\Delta\hat{t}_{i+1}$ for the exposure time of the next frame Δt_{i+1}:

1. A specific pixel must be properly exposed in at least one of the subsequent exposures.
2. At least a certain percentage of pixels must be properly exposed in both of the subsequent frames.

When these conditions are met, the output HDR is expected to have a dynamic range as high as possible without causing severe ghosting artifacts. Because in order to recover a poorly exposed region from other exposures, at least one of them must be well-exposed (condition 1) and in order to find a reliable correspondence to prevent ghosting between different exposures, at least some of the pixels must be well-exposed in multiple frames (condition 2) because image registration algorithms mostly fail in poorly exposed regions.

The first condition is satisfied by establishing an upper limit on Δt_{i+1}:

$$f^{-1}(Z^{\text{low}})/\Delta t_i \leqslant f^{-1}(Z^{\text{high}})/\Delta t_{i+1}, \tag{43}$$

$$\Delta t_{i+1} \leqslant \Delta\bar{t}_{i+1} = f^{-1}(Z^{\text{high}})\Delta t_i / f^{-1}(Z^{\text{low}}), \tag{44}$$

where $[Z^{\text{low}}, Z^{\text{high}}]$ is the user-defined pixel intensity range for well-exposedness, such that the pixels whose intensities are lower than Z^{low} are obtained from longer exposures, and the pixels whose intensities are higher than Z^{high} are obtained from shorter exposures.

The second condition is satisfied by applying refinements to $\Delta\hat{t}_{i+1}$ for achieving a desired upper limit on the percentage of overexposed pixels in LMRs and DMRs. The expected number of overexposed pixels in the next frame corresponds to the number of pixels in the current frame whose intensities are larger than $\hat{Z}$:

$$\hat{Z} = f(f^{-1}(Z^{\text{high}})\Delta t_i / \Delta\hat{t}_{i+1}). \tag{45}$$

The percentage of overexposed pixels in LMRs and DMRs is calculated by counting the number of pixels in the current frame whose intensities are larger than $\hat{Z}$. This operation is performed for different exposure settings provided by the camera. The exposure setting which satisfies the upper limit on Δt_{i+1} and the predefined limit on the percentage of the overexposed pixels in DMRs and LMRs is selected as the value of Δt_{i+1}. This process is used for searching for a longer exposure time and searching for a shorter exposure time is an analogous operation.

Once the whole sequence of frames is captured using the real-time adaptive exposure selection method described earlier, it is processed to produce an HDR video. Because the exposure settings of the captured frames are not periodic and the exposure times are not regularly spaced, available HDR reconstruction algorithms cannot be used without modifications. For this purpose, the authors use the method of Kalantari et al. [85] (Section 6.8) after making necessary modifications to their algorithm.

The method of Kalantari et al. [85] is extended by synthesizing a set of exposures for each frame in the sequence. The exposures in the set are found by recursively searching for the frames with shorter and longer exposure times, which have sufficient spatial overlap with the current frame. Once the exposure stack of each frame is formed, they are merged into HDR frames. The method of Debevec and Malik [4] is modified by introducing a probability of correct match term for the patch-based search of Kalantari et al. [85]. The results of the proposed algorithm are found to be better than the original method of Kalantari et al. [85], when compared using the mean structural similarity index (SSIM) [115] score.

REFERENCES

[1] B. Hoefflinger, High-Dynamic-Range (HDR) Vision, Springer-Verlag, Berlin, Heidelberg, 2007.
[2] K. Myszkowski, R. Mantiuk, G. Krawczyk, High dynamic range video, Synth. Lect. Comput. Graph. Anim. 1 (1) (2008) 1–158.
[3] Magic Lantern, Canon DSLR camera firmware, 2013, http://www.magiclantern.fm.
[4] P.E. Debevec, J. Malik, Recovering high dynamic range radiance maps from photographs, in: Proceedings of the 24th Annual Conference on Computer Graphics and Interactive Techniques, SIGGRAPH'97, ACM Press/Addison-Wesley Publishing Co., New York, NY, USA, 1997, pp. 369–378.
[5] M. Granados, B. Ajdin, M. Wand, C. Theobalt, H.-P. Seidel, H. Lensch, Optimal HDR reconstruction with linear digital cameras, in: IEEE Conference on Computer Vision and Pattern Recognition (CVPR), IEEE, 2010, pp. 215–222.
[6] A.O. Akyüz, E. Reinhard, Noise reduction in high dynamic range imaging, J. Vis. Commun. Image Represent. 18 (5) (2007) 366–376.
[7] O.T. Tursun, A.O. Akyüz, A. Erdem, E. Erdem, The state of the art in HDR deghosting: a survey and evaluation, in: Computer Graphics Forum, vol. 32 (2015) 348–362.
[8] S. Mann, C. Manders, J. Fung, Painting with looks: photographic images from video using quantimetric processing, in: Proceedings of the Tenth ACM International Conference on Multimedia, ACM, 2002, pp. 117–126.
[9] F.M. Candocia, Simultaneous homographic and comparametric alignment of multiple exposure-adjusted pictures of the same scene, IEEE Trans. Image Process. 12 (12) (2003) 1485–1494.
[10] G. Ward, Fast, robust image registration for compositing high dynamic range photographs from hand-held exposures, J. Graph. Tool 8 (2) (2003) 17–30.
[11] L. Cerman, V. Hlavac, Exposure time estimation for high dynamic range imaging with hand held camera, in: Proceedings of Computer Vision Winter Workshop, Czech Republic, Citeseer, 2006.

[12] A. Eden, M. Uyttendaele, R. Szeliski, Seamless image stitching of scenes with large motions and exposure differences, in: IEEE Computer Society Conference on Computer Vision and Pattern Recognition, vol. 2, 2006, pp. 2498–2505.
[13] M. Gevrekci, B.K. Gunturk, On geometric and photometric registration of images, in: IEEE International Conference on Acoustics, Speech and Signal Processing, vol. 1, IEEE, 2007.
[14] A. Tomaszewska, R. Mantiuk, Image registration for multi-exposure high dynamic range image acquisition, in: 15th International Conference Central Europe on Computer Graphics, Visualization and Computer Vision, 2007.
[15] A.A. Rad, L. Meylan, P. Vandewalle, S. Süsstrunk, Multidimensional image enhancement from a set of unregistered and differently exposed images, in: Electronic Imaging, International Society for Optics and Photonics, 2007.
[16] J. Im, S. Lee, J. Paik, Improved elastic registration for removing ghost artifacts in high dynamic imaging, IEEE Trans. Consum. Electron. 57 (2) (2011) 932–935.
[17] A.O. Akyüz, Photographically guided alignment for HDR images, in: Eurographics-Areas Papers, 2011, pp. 73–74.
[18] S. Yao, Robust image registration for multiple exposure high dynamic range image synthesis, in: IS&T/SPIE Electronic Imaging, International Society for Optics and Photonics, 2011.
[19] E.A. Khan, A.O. Akyüz, E. Reinhard, Ghost removal in high dynamic range images, in: IEEE International Conference on Image Processing, IEEE, 2006, pp. 2005–2008.
[20] M. Pedone, J. Heikkilä, Constrain propagation for ghost removal in high dynamic range images, in: VISAPP, 2008, pp. 36–41.
[21] M. Granados, H.-P. Seidel, H. Lensch, Background estimation from non-time sequence images, in: Proceedings of Graphics Interface, Canadian Information Processing Society, 2008, pp. 33–40.
[22] D.D. Sidibe, W. Puech, O. Strauss, et al., Ghost detection and removal in high dynamic range images, in: European Signal Processing Conference, 2009.
[23] S. Silk, J. Lang, Fast high dynamic range image deghosting for arbitrary scene motion, in: Proceedings of Graphics Interface, Canadian Information Processing Society, 2012, pp. 85–92.
[24] W. Zhang, W.-K. Cham, Gradient-directed composition of multi-exposure images, in: IEEE Conference on Computer Vision and Pattern Recognition, IEEE, 2010, pp. 530–536.
[25] W. Zhang, W.-K. Cham, Gradient-directed multiexposure composition, IEEE Trans. Image Process. 21 (4) (2012) 2318–2323.
[26] W.-C. Kao, C.-C. Hsu, L.-Y. Chen, C.-C. Kao, S.-H. Chen, Integrating image fusion and motion stabilization for capturing still images in high dynamic range scenes, IEEE Trans. Consum. Electron. 52 (3) (2006) 735–741.
[27] T. Grosch, Fast and robust high dynamic range image generation with camera and object movement, in: Vision, Modeling and Visualization, RWTH Aachen, 2006, pp. 277–284.
[28] K. Jacobs, C. Loscos, G. Ward, Automatic high-dynamic range image generation for dynamic scenes, IEEE Comput. Graph. Appl. 28 (2) (2008) 84–93.
[29] H.-Y. Lin, W.-Z. Chang, High dynamic range imaging for stereoscopic scene representation, in: 16th IEEE International Conference on Image Processing (ICIP), 2009, pp. 4305–4308.
[30] E. Reinhard, G. Ward, S. Pattanaik, P. Debevec, High Dynamic Range Imaging: Acquisition, Display and Image-Based Lighting, second ed., Morgan Kaufmann, San Francisco, 2010.
[31] F. Pece, J. Kautz, Bitmap movement detection: HDR for dynamic scenes, in: Conference on Visual Media Production (CVMP), IEEE, 2010, pp. 1–8.

[32] D.-K. Lee, R.-H. Park, S. Chang, Improved histogram based ghost removal in exposure fusion for high dynamic range images, in: IEEE 15th International Symposium on Consumer Electronics (ISCE), IEEE, 2011, pp. 586–591.
[33] O. Gallo, N. Gelfand, W.-C. Chen, M. Tico, K. Pulli, Artifact-free high dynamic range imaging, in: IEEE International Conference on Computational Photography (ICCP), IEEE, 2009, pp. 1–7.
[34] T.-H. Min, R.-H. Park, S. Chang, Noise reduction in high dynamic range images, Signal Image Video Process. 5 (3) (2011) 315–328.
[35] T.-H. Min, R.-H. Park, S. Chang, Histogram based ghost removal in high dynamic range images, in: IEEE International Conference on Multimedia and Expo, IEEE, 2009, pp. 530–533.
[36] Y.-S. Moon, Y.-M. Tai, J.H. Cha, S.-H. Lee, A simple ghost-free exposure fusion for embedded HDR imaging, in: IEEE International Conference on Consumer Electronics (ICCE), IEEE, 2012, pp. 9–10.
[37] S. Raman, V. Kumar, S. Chaudhuri, Blind de-ghosting for automatic multi-exposure compositing, in: SIGGRAPH ASIA Posters, ACM, 2009, p. 44.
[38] A. Srikantha, D. Sidibé, F. Mériaudeau, An SVD-based approach for ghost detection and removal in high dynamic range images, in: 21st International Conference on Pattern Recognition (ICPR), IEEE, 2012, pp. 380–383.
[39] S. Raman, S. Chaudhuri, Bottom-up segmentation for ghost-free reconstruction of a dynamic scene from multi-exposure images, in: Proceedings of the Seventh Indian Conference on Computer Vision, Graphics and Image Processing, ACM, 2010, pp. 56–63.
[40] W. Zhang, W.-K. Cham, Reference-guided exposure fusion in dynamic scenes, J. Vis. Commun. Image Represent. 23 (3) (2012) 467–475.
[41] Z. Li, S. Rahardja, Z. Zhu, S. Xie, S. Wu, Movement detection for the synthesis of high dynamic range images, in: 17th IEEE International Conference on Image Processing (ICIP), IEEE, 2010, pp. 3133–3136.
[42] T.H. Oh, J.-Y. Lee, I.-S. Kweon, High dynamic range imaging by a rank-1 constraint, in: ICIP, 2013, pp. 790–794.
[43] S. Wu, S. Xie, S. Rahardja, Z. Li, A robust and fast anti-ghosting algorithm for high dynamic range imaging, in: 17th IEEE International Conference on Image Processing (ICIP), IEEE, 2010, pp. 397–400.
[44] H.-S. Sung, R.-H. Park, D.-K. Lee, S. Chang, Feature based ghost removal in high dynamic range imaging, Int. J. Comput. Graph. Anim. 3 (4) (2013).
[45] Y.S. Heo, K.M. Lee, S.U. Lee, Y. Moon, J. Cha, Ghost-free high dynamic range imaging, in: Computer Vision–ACCV, Springer, Berlin, Heidelberg, 2010, pp. 486–500.
[46] M. Granados, K.I. Kim, J. Tompkin, C. Theobalt, Automatic noise modeling for ghost-free HDR reconstruction, ACM Trans. Graph. 32 (6) (2013) 201.
[47] J. An, S.H. Lee, J.G. Kuk, N.I. Cho, A multi-exposure image fusion algorithm without ghost effect, in: IEEE International Conference on Acoustics, Speech and Signal Processing (ICASSP), IEEE, 2011, pp. 1565–1568.
[48] C. Wang, C. Tu, An exposure fusion approach without ghost for dynamic scenes, in: 6th International Congress on Image and Signal Processing, vol. 2, IEEE, 2013, pp. 904–909.
[49] S. Raman, S. Chaudhuri, Reconstruction of high contrast images for dynamic scenes, Vis. Comput. 27 (12) (2011) 1099–1114.
[50] C. Lee, Y. Li, V. Monga, Ghost-free high dynamic range imaging via rank minimization, IEEE Signal Process. Lett. 21 (9) (2014) 1045–1049.
[51] L. Bogoni, Extending dynamic range of monochrome and color images through fusion, in: 15th International Conference on Pattern Recognition, vol. 3, IEEE, 2000, pp. 7–12.

[52] I. Hossain, B.K. Gunturk, High dynamic range imaging of non-static scenes, in: IS&T/SPIE Electronic Imaging, International Society for Optics and Photonics, 2011.
[53] H. Zimmer, A. Bruhn, J. Weickert, Freehand HDR imaging of moving scenes with simultaneous resolution enhancement, Comput. Graph. Forum 30 (2) (2011) 405–414.
[54] S. Ferradans, M. Bertalmio, E. Provenzi, V. Caselles, Generation of HDR images in non-static conditions based on gradient fusion, in: VISAPP, 2012, pp. 31–37.
[55] T. Jinno, M. Okuda, Multiple exposure fusion for high dynamic range image acquisition, IEEE Trans. Image Process. 21 (1) (2012) 358–365.
[56] D. Hafner, O. Demetz, J. Weickert, Simultaneous HDR and optic flow computation, in: Proceedings of 22nd International Conference on Pattern Recognition (ICPR 2014), IEEE, 2014.
[57] N. Menzel, M. Guthe, Freehand HDR photography with motion compensation, in: VMV, 2007, pp. 127–134.
[58] S.-C. Park, H.-H. Oh, J.-H. Kwon, W. Choe, S.-D. Lee, Motion artifact-free HDR imaging under dynamic environments, in: 18th IEEE International Conference on Image Processing (ICIP), IEEE, 2011, pp. 353–356.
[59] J. Zheng, Z. Li, Z. Zhu, S. Rahardja, A hybrid patching scheme for high dynamic range imaging, in: Asia Pacific Signal and Information Processing Association Annual Summit and Conference, 2011.
[60] R.R. Orozco, I. Martín, C. Loscos, P.-P. Vasquez, Full high-dynamic range images for dynamic scenes, in: SPIE Photonics Europe, International Society for Optics and Photonics, 2012.
[61] J. Hu, O. Gallo, K. Pulli, Exposure stacks of live scenes with hand-held cameras, in: Computer Vision–ECCV, Springer, Berlin, Heidelberg, 2012, pp. 499–512.
[62] P. Sen, N.K. Kalantari, M. Yaesoubi, S. Darabi, D.B. Goldman, E. Shechtman, Robust patch-based HDR reconstruction of dynamic scenes, ACM Trans. Graph. 31 (6) (2012) 203.
[63] J. Hu, O. Gallo, K. Pulli, X. Sun, HDR deghosting: how to deal with saturation, in: IEEE Conference on Computer Vision and Pattern Recognition (CVPR), IEEE, 2013.
[64] J. Zheng, Z. Li, Z. Zhu, S. Wu, S. Rahardja, Patching of moving objects for ghosting-free HDR synthesis, in: ACM SIGGRAPH Posters, ACM, 2012, p. 62.
[65] J. Zheng, Z. Li, Z. Zhu, S. Wu, S. Rahardja, Hybrid patching for a sequence of differently exposed images with moving objects, IEEE Trans. Image Process. 22 (12) (2013) 5190–5201.
[66] O.T. Tursun, A.O. Akyüz, A. Erdem, E. Erdem, An objective deghosting quality metric for HDR images, Comput. Graph. Forum 35 (2) (2016) 139–152.
[67] M. Aggarwal, N. Ahuja, Split aperture imaging for high dynamic range, Int. J. Comput. Vis. 58 (1) (2004) 7–17.
[68] H. Wang, R. Raskar, N. Ahuja, High dynamic range video using split aperture camera, in: IEEE 6th Workshop on Omnidirectional Vision, Camera Networks and Non-classical Cameras, Washington, DC, USA, 2005.
[69] M.D. Tocci, C. Kiser, N. Tocci, P. Sen, A versatile HDR video production system, in: ACM Transactions on Graphics (TOG), vol. 30, ACM, 2011, p. 41.
[70] J. Kronander, S. Gustavson, J. Unger, Real-time HDR video reconstruction for multi-sensor systems, in: ACM SIGGRAPH 2012 Posters, ACM, 2012, p. 65.
[71] J. Kronander, S. Gustavson, G. Bonnet, J. Unger, Unified HDR reconstruction from raw CFA data, in: 2013 IEEE International Conference on Computational Photography (ICCP), IEEE, 2013, pp. 1–9.

[72] V. Ramachandra, M. Zwicker, T. Nguyen, HDR imaging from differently exposed multiview videos, in: 3DTV Conference: The True Vision – Capture, Transmission and Display of 3D Video, 2008, pp. 85–88.

[73] J. Bonnard, C. Loscos, G. Valette, J.-M. Nourrit, L. Lucas, High-dynamic range video acquisition with a multiview camera, in: SPIE Photonics Europe, International Society for Optics and Photonics, 2012, p. 84360A.

[74] E. Selmanovic, K. Debattista, T. Bashford-Rogers, A. Chalmers, Enabling stereoscopic high dynamic range video, Signal Process. Image Commun. 29 (2) (2014) 216–228.

[75] M. Bätz, T. Richter, J.-U. Garbas, A. Papst, J. Seiler, A. Kaup, High dynamic range video reconstruction from a stereo camera setup, Signal Process. Image Commun. 29 (2) (2014) 191–202.

[76] M. Schoberl, A. Belz, J. Seiler, S. Foessel, A. Kaup, High dynamic range video by spatially non-regular optical filtering, in: 2012 19th IEEE International Conference on Image Processing, 2012.

[77] M. Niskanen, View dependent enhancement of the dynamic range of video, in: 18th International Conference on Pattern Recognition (ICPR 2006), vol. 1, IEEE, 2006, pp. 984–987.

[78] B. Guthier, S. Kopf, W. Effelsberg, Histogram-based image registration for real-time high dynamic range videos, in: 2010 17th IEEE International Conference on Image Processing (ICIP), IEEE, 2010, pp. 145–148.

[79] B. Guthier, S. Kopf, W. Effelsberg, A real-time system for capturing HDR videos, in: Proceedings of the 20th ACM International Conference on Multimedia, ACM, 2012, pp. 1473–1476.

[80] S.B. Kang, M. Uyttendaele, S. Winder, R. Szeliski, ACM, High dynamic range video, ACM Trans. Graph. 22 (3) (2003) 319–325.

[81] T. Castro, A. Chapiro, M. Cicconet, L. Velho, Towards mobile HDR video, in: Proceedings of the Eurographics-Areas Papers, 2011, pp. 75–76.

[82] P. Sand, S. Teller, Video matching, ACM Trans. Graph. 23 (3) (2004) 592–599.

[83] A. Chapiro, M. Cicconet, L. Velho, Filter based deghosting for exposure fusion video, in: ACM SIGGRAPH Posters, ACM, 2011, p. 33.

[84] L. Velho, Histogram-based HDR video, in: SIGGRAPH, vol. 7, 2007, p. 62.

[85] N.K. Kalantari, E. Shechtman, C. Barnes, S. Darabi, D.B. Goldman, P. Sen, Patch-based high dynamic range video, ACM Trans. Graph. 32 (6) (2013) 202.

[86] S. Mangiat, J. Gibson, High dynamic range video with ghost removal, in: SPIE Optical Engineering+ Applications, International Society for Optics and Photonics, 2010.

[87] Y. Li, C. Lee, V. Monga, A MAP estimation framework for HDR video synthesis, in: IEEE International Conference on Image Processing (ICIP), 2015.

[88] S. Mangiat, J. Gibson, Spatially adaptive filtering for registration artifact removal in HDR video, in: 2011 18th IEEE International Conference on Image Processing (ICIP), IEEE, 2011, pp. 1317–1320.

[89] M. Aggarwal, N. Ahuja, High dynamic range panoramic imaging, in: Proceedings of Eighth IEEE International Conference on Computer Vision, 2001 (ICCV 2001), vol. 1, 2001, pp. 2–9.

[90] J. Prevoteau, S. Chalençon-Piotin, D. Debons, L. Lucas, Y. Remion, Multiview shooting geometry for multiscopic rendering with controlled distortion, Int. J. Digit. Multimed. Broadcast. 2010 (2010) 11.

[91] C. Niquin, S. Prévost, Y. Remion, An occlusion approach with consistency constraint for multiscopic depth extraction, Int. J. Digit. Multimed. Broadcast. 2010 (2010) 1–8.

[92] A. Chalmers, G. Bonnet, F. Banterle, P. Dubla, K. Debattista, A. Artusi, C. Moir, High-dynamic-range video solution, in: ACM SIGGRAPH ASIA 2009 Art Gallery & Emerging Technologies: Adaptation (SIGGRAPH ASIA'09), ACM, New York, NY, USA, 2009, pp. 71–71.
[93] S.K. Nayar, T. Mitsunaga, High dynamic range imaging: spatially varying pixel exposures, in: IEEE Conference on Computer Vision and Pattern Recognition, vol. 1, IEEE, 2000, pp. 472–479.
[94] S. Kavusi, A. El Gamal, Quantitative study of high-dynamic-range image sensor architectures, in: Electronic Imaging 2004, International Society for Optics and Photonics, 2004, pp. 264–275.
[95] Q. Zhou, S. Guo, G. Du, Y. Wang, Y. Chang, High-dynamic-range photodetecting scheme based on PEPT with a large output swing, IEEE Trans. Electron Dev. 59 (5) (2012) 1423–1429.
[96] A. Serrano, F. Heide, D. Gutierrez, G. Wetzstein, B. Masia, Convolutional sparse coding for high dynamic range imaging, Comput. Graph. Forum 35 (35) (2016) 153–163.
[97] D.G. Lowe, Distinctive image features from scale-invariant keypoints, Int. J. Comput. Vis. 60 (2) (2004) 91–110.
[98] M.A. Fischler, R.C. Bolles, Random sample consensus: a paradigm for model fitting with applications to image analysis and automated cartography, Commun. ACM 24 (6) (1981) 381–395.
[99] R.E. Kalman, A new approach to linear filtering and prediction problems, J. Fluids Eng. 82 (1) (1960) 35–45.
[100] B. Guthier, S. Kopf, W. Effelsberg, Capturing high dynamic range images with partial re-exposures, in: 2008 IEEE 10th Workshop on Multimedia Signal Processing, 2008, pp. 241–246.
[101] B. Guthier, S. Kopf, W. Effelsberg, Optimal shutter speed sequences for real-time HDR video, in: 2012 IEEE International Conference on Imaging Systems and Techniques (IST), 2012, pp. 303–308.
[102] B.D. Lucas, T. Kanade, et al., An iterative image registration technique with an application to stereo vision, in: IJCAI, vol. 81, 1981, pp. 674–679.
[103] J.R. Bergen, P. Anandan, K.J. Hanna, R. Hingorani, Hierarchical model-based motion estimation, in: Computer Vision (ECCV'92), Springer, Berlin, Heidelberg, 1992, pp. 237–252.
[104] T. Mitsunaga, S.K. Nayar, Radiometric self calibration, in: IEEE Computer Society Conference on Computer Vision and Pattern Recognition, vol. 1, IEEE, 1999.
[105] E. Reinhard, M. Stark, P. Shirley, J. Ferwerda, ACM, Photographic tone reproduction for digital images, ACM Trans. Graph. 21 (3) (2002) 267–276.
[106] C. Harris, M. Stephens, A combined corner and edge detector, in: Alvey vision Conference, Manchester, UK, vol. 15, 1988, p. 50.
[107] C.G. Atkeson, A.W. Moore, S. Schaal, Locally weighted learning for control, in: Lazy Learning, Springer, New York, 1997, pp. 75–113.
[108] J. Shi, C. Tomasi, Good features to track, in: 1994 IEEE Computer Society Conference on Computer Vision and Pattern Recognition, 1994 (Proceedings CVPR'94), IEEE, 1994, pp. 593–600.
[109] P. Sand, S. Teller, Video Matching, Technical Report, MIT, 2004.
[110] S.-J. Youm, W.-H. Cho, K.-S. Hong, High dynamic range video through fusion of exposured-controlled frames, in: MVA, 2005, pp. 546–549.
[111] P.J. Burt, E.H. Adelson, A multiresolution spline with application to image mosaics, ACM Trans. Graph. (TOG) 2 (4) (1983) 217–236.
[112] S.K. Nayar, V. Branzoi, Adaptive dynamic range imaging: optical control of pixel exposures over space and time, in: 9th IEEE International Conference on Computer Vision, IEEE, 2003, pp. 1168–1175.

[113] A.M. Tourapis, Enhanced predictive zonal search for single and multiple frame motion estimation, in: Electronic Imaging 2002, International Society for Optics and Photonics, 2002, pp. 1069–1079.
[114] S. Paris, F. Durand, A fast approximation of the bilateral filter using a signal processing approach, in: Computer Vision–ECCV 2006, Springer, Heidelberg, 2006, pp. 568–580.
[115] Z. Wang, A.C. Bovik, H.R. Sheikh, E.P. Simoncelli, Image quality assessment: from error visibility to structural similarity, IEEE Trans. Image Process. 13 (4) (2004) 600–612.
[116] T. Mertens, J. Kautz, F. Van Reeth, Exposure fusion, in: 15th Pacific Conference on Computer Graphics and Applications, IEEE, 2007, pp. 382–390.
[117] D. Simakov, Y. Caspi, E. Shechtman, M. Irani, Summarizing visual data using bidirectional similarity, in: IEEE Conference on Computer Vision and Pattern Recognition, IEEE, 2008, pp. 1–8.
[118] T.-H. Oh, J.-Y. Lee, Y.-W. Tai, I.S. Kweon, Robust high dynamic range imaging by rank minimization, IEEE Trans. Pattern Anal. Mach. Intell. 37 (6) (2015) 1219–1232.
[119] C. Lee, Y. Li, V. Monga, Ghost-free high dynamic range imaging via rank minimization, IEEE Signal Process. Lett. 21 (9) (2014) 1045–1049.
[120] Y. Gryaditskaya, T. Pouli, E. Reinhard, K. Myszkowski, H.-P. Seidel, Motion aware exposure bracketing for HDR video, in: Computer Graphics Forum, vol. 34, Wiley Online Library, 2015, pp. 119–130.
[121] F. Urban, J.-F. Nezan, M. Raulet, HDS, a real-time multi-DSP motion estimator for MPEG-4 H. 264 AVC high definition video encoding, J. Real-Time Image Process. 4 (1) (2009) 23–31.

CHAPTER 2

Creating HDR Video Using Retargetting

F. Banterle*, J. Unger†
*CNR-ISTI, Pisa, Italy
†Linköping University, Linköping, Sweden

Contents

1. INTRODUCTION

Capture of HDR still images is now well known and from many aspects, a solved problem; see [1, 2] for an overview. However, HDR video capture still remains a challenge. There are a number of professional systems, such as the Arri Alexa XT and the Red Epic Dragon, with an extended dynamic range of up to 14–16.5 *f-stops*, and research prototypes [3, 4], with a dynamic range of up to 20–24 *f-stops*. However, these camera systems either still have a limited dynamic range and are very costly (commercial systems), or have not matured enough to reach consumer markets (research prototypes). An important area is to develop methods that allow users to capture HDR video using low cost off-the-shelf technology, such as ordinary video cameras and standard DSLR cameras.

High Dynamic Range Video
http://dx.doi.org/10.1016/B978-0-12-809477-8.00002-9

This chapter gives an overview of two practical methods for generating HDR videos by retargeting information from a set of sparsely captured HDR images or panoramas onto a standard dynamic range (SDR) videos, captured in the same scene. The first method by Bhat et al. [5] aims to register and fuse a dense SDR video with sparse HDR photographs in static scenes with moving cameras by exploiting computer vision algorithms for registration. The second method by Banterle et al. [6] solves the problem in the case of a static SDR video camera (a static background) and a dynamic scene, such as people or objects moving in the scene. The method is carried out as a completely automatic postprocessing phase, i.e., after the input footage has been captured. The two techniques are useful in different application scenarios: from the acquisition of medium-to-high quality videos, to the typical case of webcams, where a satisfying trade-off in exposure between the subject in foreground and the usually bright background is difficult (if not impossible) to find.

2. BACKGROUND IN HDR VIDEO AND IMAGE RETARGETING

The goal of the two methods described in this chapter is to create HDR videos by fusing or retargeting image data captured as dense SDR video footage, and sparse image data captured using HDR imaging techniques. We start this chapter by giving an overview of methods for HDR capture, and an overview of related work in, and the concept of, image retargeting.

2.1 HDR Video Capture

The dynamic range of a camera refers to the ratio between the highest intensity and the lowest intensity and the sensor can accurately capture within the same frame. The lowest intensity is usually defined by an acceptable signal-to-noise ratio. Here, we use the term HDR image to mean an image, which covers the full dynamic range of the scene. This means that no pixels are saturated and that all pixels can be trusted as a measurement of the scene radiance with a linear response to the number of photons detected. Each pixel and color channel in an HDR image is usually represented as a 32- or 16-bit precision floating point number, instead of an 8-bit integer as in traditional images.

The typical dynamic range exhibited by current high-end and consumer camera systems is in the order of 10,000:1. This should be compared to the dynamic range found in real scenes, which in many cases extends beyond 1,000,000:1, from direct light sources to shadows. This is significantly more than the dynamic range of most camera systems can capture. Fig. 1 shows an

Fig. 1 Four gamma mapped exposures three *f-stops* apart created from a 32-bit HDR image. The virtual exposures are generated by scaling the floating point pixel values in the original HDR image and performing gamma mapping and 8-bit quantization.

example of HDR image as four tone mapped exposures. The four images are generated three *f-stops* apart from the original HDR image by linear scaling of the input pixels, gamma mapping, and finally quantization of the floating point data into 8 bits per pixel and (R,G,B) color channel.

In the professional segment there are cameras, e.g., the Red Epic Dragon and Arri Alexa XT, with an extended dynamic range of up to 16 *f-stops*, corresponding to a dynamic range in the order of 100,000:1. There are also image sensors with a logarithmic response that can capture a significantly higher dynamic range. However, these sensors are, in general, not suitable for high-quality imaging applications because they usually have a comparably low resolution and suffer from image noise in darker regions. To enable capture of images with a dynamic range larger than the approximately 16 *f-stops* offered by high-end cameras, it is currently necessary to perform some form of multiplexing. The multiplexing can be carried out in different ways, e.g., in the temporal domain by capturing different exposures, in the radiometric domain by using multiple image sensors, or in the spatial domain by inserting filters that vary the transmittance of light on a per-pixel level.

The traditional approach for capturing HDR images is to capture a set of different exposures covering the dynamic range of the scene and merge them into the final HDR image [7–9]. This approach is often referred to as *exposure bracketing*, and many modern cameras can be set to do this automatically. The main drawback with this method is that it requires both the camera and the scene to be kept static during capture, due to the fact that a set of images with different exposure settings needs to be captured. Unger et al. [10, 11] developed an HDR video camera system where the time disparity between the different exposures was minimized by capturing four different exposures back-to-back for each row on the sensor in a rolling shutter fashion, instead of waiting for the entire image to be captured. To extend the dynamic range further, they also A/D converted each pixel value twice with different amplification of the analog pixel readout, i.e., the images were captured with two different ISO values simultaneously. There are also algorithms for compensating for the misalignments introduced by camera motion or dynamic objects [12]. However, these methods are, in most cases, not fully robust and may still lead to artifacts.

Another approach is to use multiple, synchronized camera sensors imaging the scene through the same optical system. This technique is suitable for capture of HDR video, because the sensors capture all exposures at the same time. Robustness to dynamic scenes and correct motion blur can be ensured by using a common exposure time. The exposures captured by the sensors can be varied by inserting different filters. For example, Kronander et al. [4, 13] proposed natural density (ND) filters in front of four different sensors built into the same camera setup. Similarly, Froehlich et al. [14] used two Arri Alexa digital film cameras capturing the scene through zero baseline stereo setup. However, the use of ND filters in the optical path leads to a waste of light, i.e., longer exposure times. This is because a lot of the incident radiance is simply filtered out. To avoid this issue, Tocci et al. [3] introduced a multisensor setup with three sensors without ND-filters. Instead of using ND filters, they show how the beam-splitters themselves can be used to reflect different amounts of the incident light onto the different sensors.

Another possibility is to trade spatial resolution against dynamic range by using a sensor where the response varies between pixels. The advantage is that all exposures are captured at the same time and with the same exposure time. Spatially varying exposures [15] can be achieved by placing a filter mask with different transmittance on top of the sensor. This idea is similar to color imaging using a single sensor using, e.g., a Bayer pattern color filter array, but with the difference that in addition to the (R,G,B) color filters, there are also ND filters distributed over the sensor. Another approach is to let the per-pixel gain, or ISO, vary between groups of pixels over the sensor; see for example Hajisharif et al.'s work [16]. This functionality is currently available for off-the-shelf Canon cameras running the Magic Lantern firmware[1] as further described in Chapter 4. The disadvantage of the spatial multiplexing approach is the trade-off between resolution and achievable dynamic range, as too many different exposures may lead to a lower resolution in the reconstructed HDR image.

2.2 Image Retargeting

Image retargeting, in the sense of improving existing images or videos with higher quality content, is a very broad research topic in computer graphics and imaging. Retargeting can happen for different image/video frame attributes, such as image resolution, color information, temporal resolution, and dynamic range; see [17] for an overview. In this section, we want to highlight a few works which are closely related to the problem of enhancing SDR content using HDR reference photographs.

Wang et al. [18] proposed a method for increasing the dynamic range and transferring the detail from a source LDR image into a target LDR image. While the dynamic range expansion is automatic by fitting Gaussian profiles around overexposed regions, the details transfer part is manual. In this case, a user needs to transfer high-frequency details from a well-exposed area (or another image) to an overexposed or an underexposed area using a tool similar to the *healing tool* of Adobe Photoshop.[2] Although this method produces high-quality results, it cannot be applied to videos because the user interaction is extremely heavy.

Regarding retargeting videos resolution, Ancuti [19] proposed a simple technique to transfer detail from a high-resolution image to the video. In their work, SIFT features [20] are extracted from both high-resolution images and low-resolution video frames, and then they are matched. Matched patches in the high-resolution photographs are then copied onto the low-resolution frames, obtaining a high-resolution output video. Similarly, Gupta [21] extended this concept by designing a framework based on high-quality optical flow and image-based rendering. This can be implemented in video-cameras that can acquire both photographs and videos. However, both these works do not handle HDR information during the retargeting of the input SDR video.

[1] http://www.magiclantern.fm.
[2] http://www.adobe.com/products/photoshop.html.

3. DYNAMIC CAMERA AND STATIC SCENE

SDR videos can be enhanced when the scene is static, but the camera moves using Bhat et al.'s system [5]. This system, depicted in Fig. 2, requires as input a few HDR photographs of a static scene and an SDR video of the scene, and it outputs an enhanced video where the HDR information from photographs is transferred onto the SDR video.

The system has two main components:

- a geometry estimation component for estimating depths and correspondences;
- an image-based renderer for video reconstruction.

3.1 Geometry Estimation

As the first step, the system applies a structure-from-motion (SfM) algorithm [22] in order to compute a sparse point cloud, including projection matrices for each photograph and video frame, and a list of the viewpoints from which each scene point is visible (see Fig. 3).

Then, a modified multiview stereo (MVS) algorithm [23] is employed for computing depth maps for each video frame and HDR photograph. This MVS algorithm segments the image based on colors, and it then

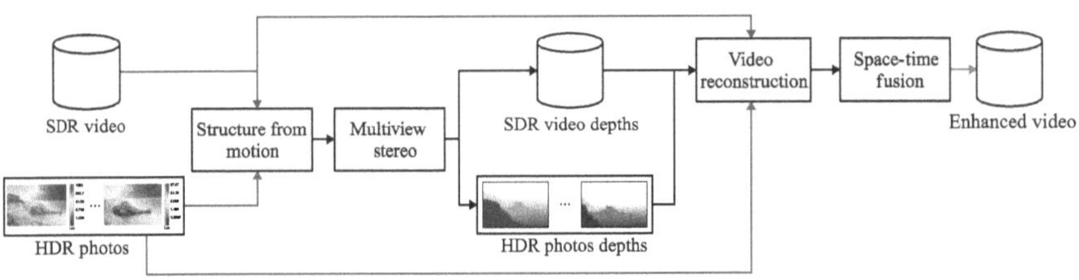

Fig. 2 Bhat et al. [5] pipeline for enhancing SDR videos.

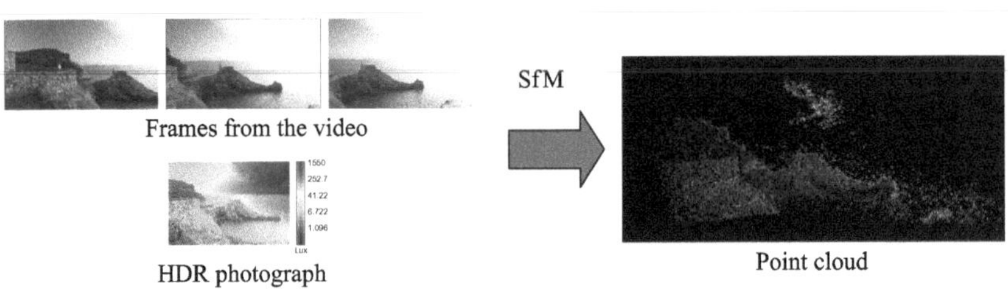

Fig. 3 An example of SfM applied to an SDR video and HDR photographs. *(The original images and videos are courtesy of T. Sergej.)*

Fig. 4 An example of depth maps generated by an MVS algorithm. At the top, the color HDR photograph (*tone mapped*) and SDR video frames. At the bottom, the reconstructed depth maps. *(The original images and videos are courtesy of T. Sergej.)*

computes disparity for each segment by constructing a pair-wise Markov random field (MRF) for each image. Bhat et al. extended it to take into account heterogeneous datasets (photographs and frames), wide range of disparity planes, and 3D points in the point cloud from SfM to improve depth estimation. An example of generated depth maps is shown in Fig. 4.

3.2 Video Reconstruction

The correspondence found using SfM and the MVS modified algorithm (see previous section) are then used to reconstruct an HDR video, using the HDR information of the photographs while preserving the temporal dynamics of the SDR input video.

An approach to solve this problem is to use the classic view interpolation approach [24, 25]. However, images warped with view interpolation can suffer in ghosting artifacts and loss of high-frequency details. To overcome these issues, the authors proposed to formulate the reconstruction problem as a labeling problem in a MRF network. The goal is to assign to each pixel p of the ith video frame, V_i, a label, $L(P)$, which indicates the candidate HDR photograph that should contribute to the final result. This candidate is chosen among the N nearest HDR photographs to V_i that are reprojected from the viewpoint of V_i. This leads to the following cost function:

$$C(L) = \sum_{p \in V_i} C_D(p, L(p)) + \lambda \sum_{\{p,q\} \in n} C_S(p, q, L(p), L(q)), \qquad (1)$$

where C_D is the data cost function encouraging video pixels to be reconstructed from photographs with similar color and depth, C_S is the smoothness term similar to the one by Kwatra et al. [26], n is a set of all eight-connected neighbors in V_i, p and q are neighboring pixels defined by N, and λ is a trade-off between C_D and C_S. Eq. (1) is solved using a graph-cut optimization [27] obtaining a reconstructed video R.

The output video computed in the MRF reconstruction may suffer from artifacts such as lack of rich temporal variations, seams in the frame (i.e., different HDR photographs used in different areas of the frame), holes (due to missing portions of the scene not seen in any of the HDR photographs). To solve these issues, a gradient domain composition is employed by using motion compensated temporal gradients, G_t, from the input SDR video, and spatial gradients, G_x and G_y, from R. Note that in the case of holes in R; temporal gradients from the original videos are copied and matched using color transfer [28]. The final enhanced video, E, is created by solving an over constrained linear system defined by the following constraints in a gradient domain fashion as:

$$E(x+1, y, t) - E(x, y, t) = G_x(x, y, t), \tag{2}$$

$$E(x, y+1, t) - E(x, y, t) = G_y(x, y, t), \tag{3}$$

$$E(x+u, y+v, t) - E(x, y, t) = G_t(x, y, t), \tag{4}$$

where (u, v) is a motion vector linking the pixel at (x, y, t) to its corresponding one in the frame $t+1$. Eq. (4) can be solved with a conjugate gradient solver [29]. Note that Eq. (4) requires to have all the videos to be in memory. If this is not the case, the linear system can be solved for slabs of 20–30 frames using Dirichlet boundary conditions.

3.3 Discussion

Bhat et al. successfully tested for retargeting SDR videos (see Fig. 5). Furthermore, this system is extremely flexible, and it can be employed for different tasks such as super-resolution, exposure correction, video editing, camera shake removal, and object removal.

One main drawback of the system is still the presence of some image artifacts in the output video, due to the errors of some computer vision algorithms used in the pipeline, such as oversegmentation in the MVS, and imprecise projection matrices from SfM.

Fig. 5 An example of the final HDR frames reconstruction. At the top, the input SDR video frames. At the bottom, the same frames after HDR reconstruction and tone mapping. *(The original images and videos are courtesy of T. Sergej.)*

Timing Bhat et al. implemented the system in C++ without optimizations, reporting on average around 5 minutes for enhancing an 853 × 480 frame on an unspecified 2007 machine.

4. STATIC CAMERA AND DYNAMIC SCENE

When the camera is static and the scene is dynamic, Banterle et al.'s algorithm [6] can be employed. This method, depicted in Fig. 6, requires as input a single HDR photograph of the background and an SDR video of the scene taken using a static camera, and it outputs an enhanced video where the HDR information from HDR background photograph

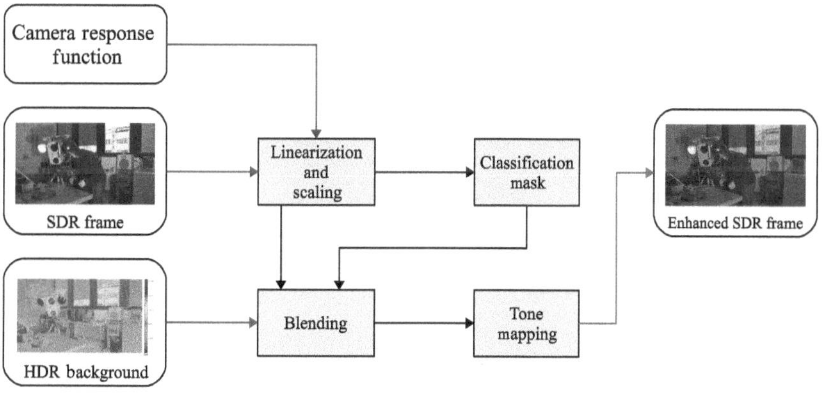

Fig. 6 Banterle et al. [6] pipeline for augmenting SDR videos.

is transferred onto the SDR video. Note that the exposure of the input SDR video footage is manually set in order to have actors or important moving objects in the scene well exposed.

4.1 The Blending Algorithm

The HDR background photograph and the SDR video are blended together in a straightforward way. As the first step, the SDR video footage is linearized applying the inverse camera response function of the SDR video-camera. After linearization, the SDR video is scaled by the capturing shutter speed, obtaining a normalized SDR video with absolute values. Linearization and scaling are important in order to match intensities and colors with the HDR background photograph. In fact, this makes it possible to use less computationally expensive techniques for the blending stage.

At this point, the HDR image and the normalized and scaled SDR frame are linearly blended in the logarithm domain, using a selection mask M, which classifies background and actors. The blending is applied in the logarithmic domain to avoid seams at the mask's boundaries. Then, the blended image is exponentiated to obtain the final radiance map. This straightforward blend is enough to obtain no seams or other kind of artifacts. Other techniques, as Laplacian pyramids [30] and Gradient Domain editing [31–33], produce similar results of the linear blending at higher computational costs. Furthermore, in some cases, the colors are slightly shifted; see Fig. 7 for a comparison.

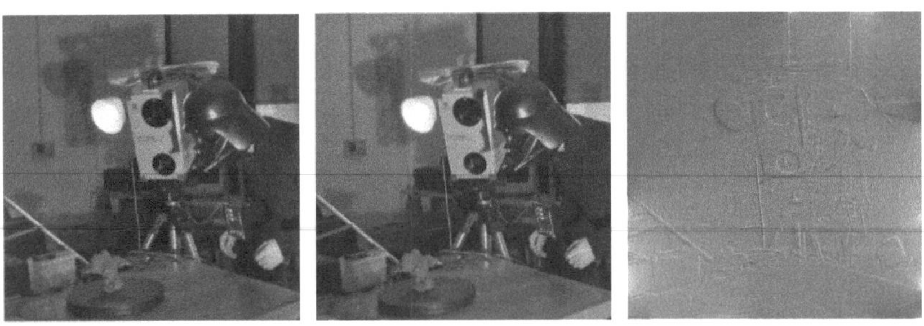

(A) (B) (C)

Fig. 7 An example of different blending algorithms: (A) Blending in the gradient domain. (B) Blending in the spatial domain. (C) The difference between (A) and (B). Note that gradients are slightly more enhanced in (A) and colors are slightly shifted toward a bluish tint (light gray in print versions).

4.2 The Classification Mask

The classification mask, M, is computed using thresholding on overexposed and underexposed pixel values on the luminance channel for each frame of the video. Authors found that 0.95 and 0.05 are, respectively, good threshold for overexposed and underexposed thresholds; using normalized RGB values. Thresholding can produce groups of single pixels in the image, which is typically to be considered as noise. Therefore, morphological operators, erosion followed by dilation, need to be applied. Typically, 3–5 iterations are enough for obtaining high-quality results on full HD content (1920×1080). Finally, the mask is cross bilateral filtered with the original SDR luminance frame using a fast bilateral filter [34, 35] ($\sigma_s = 16$ and $\sigma_r = 0.1$ for full HD content) in order to smoothly extend the classification to strong edges. Fig. 8 shows an example of the different steps for calculating the mask.

4.3 Discussion

This algorithm has a very simple acquisition step, and the processing is fully automatic. Therefore, it can be used for different setups, such as webcams and a medium/high-end cameras. Below, we give an overview of different use cases using input SDR videos of different quality.

Webcams. Webcams are usually not able to find a trade-off exposure between the foreground subject and the background, especially during video calls. To test the method, Banterle et al. used a Logitech QuickCam Pro 9000, which captures videos with a 1280×720 resolution. The example in Fig. 9 shows the selection of the best exposure for the main subject, while background information is recovered during the blending. Fig. 9C shows that a convincing HDR can be obtained with this device, although part of the range is missing due the limitations of the camera.

(A) (B) (C)

Fig. 8 An example of the classification mask: (A) A frame from an SDR video. (B) The application of thresholding to (A). (C) The final mask after applying the bilateral filter.

(A) (B) (C)

Fig. 9 A frame of a short sequence taken with a Logitech QuickCam Pro 9000 (1280 × 720 resolution): (A) A tone mapped version of the HDR reference of the background. (B) A frame from the original SDR video. (C) The same frame in (B) after enhancement and tone mapping.

(A) (B) (C)

Fig. 10 A frame of a video taken with Canon 550D (1920 × 1080 resolution): (A) A tone mapped version of the HDR reference of the background. (B) A frame from the original SDR video. (C) The same frame in (B) after enhancement and tone mapping.

Medium/High-End Cameras. Banterle et al. also evaluated the method using input SDR video from a Canon 550D, which is able to acquire videos at 1920 × 1080 resolution. The method can work in difficult lighting conditions. In the example shown in Fig. 10, a tone mapped version of the HDR background is shown in Fig. 10A. Fig. 10B shows a frame of the original SDR video, where most of the background and the sky are overexposed. Using the HDR background image, the enhanced video recovers the appearance of the sky and of several overexposed parts (see Fig. 10C).

A second example, Fig. 11, shows an indoor environment, where the light and part of the outdoor scene are lost. The enhanced video recovers this information. The colors in the frame enhanced are different from the ones in the SDR frame because the linearization process matches color curves between HDR background and SDR frame.

Timing. Authors reported on average less than 9 seconds for a 1920 × 1080 frame on an Intel Core 2 Duo at 2.33 GHz equipped with 3 GB of memory and Windows 7 using a MATLAB implementation.

(A) (B)

Fig. 11 An example of indoor scene taken with a Canon 550D (1920 × 1080): (A) A frame from the original SDR video. (B) The same frame in (A) after enhancement and tone mapping.

5. SUMMARY

This chapter presented two methods for augmenting SDR video sequences with HDR information in order to create HDR videos. The goal of the two methods is to fill in saturated regions in the SDR video frames by retargeting nonsaturated image data from a sparse set of HDR images. Table 1 shows a summary of the flexibility of the two approaches as compared to using an HDR video camera system as described in Section 2.

The HDR video cameras are naturally better equipped to capture fully dynamic HDR videos. However, current HDR video cameras are either research prototypes or not able to capture more than up to around 16 f-stops. The first method, by Bhat et al. [5], uses SfM techniques to register the SDR video and the sparsely captured HDR images to the same frame of reference and replaces pixels that are saturated in the SDR video with information from the HDR images. However, the technique may fail in regions where there is no overlap between the SDR video and the HDR images. The second method, by Banterle et al. [6], works for dynamic scenes, but it assumes that the camera is static, i.e., no camera movements. Under the assumption that moving people and/or objects are well exposed in the SDR video, this method replaces saturated pixel values

Table 1 A comparison between using specialized HDR video cameras and the two retargeting methods described in this chapter

Capture type	Camera movement	Scene type
Bhat et al. [5]	Dynamic	Static
Banterle et al. [6]	Static	Dynamic
HDR native capture	Dynamic	Dynamic

in the background in order to generate an HDR video covering the full dynamic range of the scene. In conclusion, if the HDR images used to augment the SDR footage are captured carefully, the retargeting methods described in this chapter are a good option for generating HDR videos.

REFERENCES

[1] E. Reinhard, G. Ward, S. Pattanaik, P. Debevec, W. Heidrich, K. Myszkowski, High Dynamic Range Imaging: Acquisition, Display, and Image-Based Lighting, second ed., The Morgan Kaufmann Series in Computer Graphics, Morgan Kaufmann Publishers Inc., San Francisco, CA, USA, 2010.
[2] F. Banterle, A. Artusi, K. Debattista, A. Chalmers, Advanced High Dynamic Range Imaging: Theory and Practice, first ed., AK Peters, Ltd., Natick, MA, 2011.
[3] M.D. Tocci, C. Kiser, N. Tocci, P. Sen, A versatile HDR video production system, ACM Trans. Graph. (Proc. SIGGRAPH 2011) 30 (4) (2011) 41:1–41:10.
[4] J. Kronander, S. Gustavson, G. Bonnet, J. Unger, Unified HDR reconstruction from raw CFA data, in: IEEE International Conference on Computational Photography (ICCP), 2013.
[5] P. Bhat, C.L. Zitnick, N. Snavely, A. Agarwala, M. Agrawala, B. Curless, M. Cohen, S.B. Kang, Using photographs to enhance videos of a static scene, in: J. Kautz, S. Pattanaik (Eds.), Rendering Techniques 2007 (Proceedings Eurographics Symposium on Rendering), Eurographics, 2007, pp. 327–338.
[6] F. Banterle, M. Dellepiane, R. Scopigno, Enhancement of low dynamic range videos using high dynamic range backgrounds, in: R. Mantiuk, E. Reinhard (Eds.), Annex to the Conference Proceedings: Areas Papers on HDR, Eurographics Area Papers on HDR, Eurographics Association, 2011, pp. 57–62.
[7] S. Mann, R.W. Picard, Being 'undigital' with digital cameras: extending dynamic range by combining differently exposed pictures, in: Proceedings of IS&T 46th Annual Conference, 1995, pp. 422–428.
[8] M.A. Robertson, S. Borman, R.L. Stevenson, Dynamic range improvement through multiple exposures, in: IEEE International Conference on Image Processing, 1999, pp. 159–163.
[9] P. Debevec, J. Malik, Recovering high dynamic range radiance maps from photographs, in: Proceedings of the 24th Annual Conference on Computer Graphics and Interactive Techniques (SIGGRAPH'97), ACM Press/Addison-Wesley Publishing Co., New York, NY, USA, 1997, pp. 369–378.
[10] J. Unger, S. Gustavson, M. Ollila, M. Johannesson, A real time light probe, in: Eurographics, Short Papers, 2004, pp. 17–21.
[11] J. Unger, S. Gustavson, High-dynamic-range video for photometric measurement of illumination, in: SPIE, vol. 6501, 2007.
[12] O.T. Tursun, A.O. Akyuz, A. Erdem, E. Erdem, The state of the art in HDR deghosting: a survey and evaluation, Comput. Graph. Forum 34 (2) (2015) 683–707 (Proc. Eurographics).
[13] J. Kronander, S. Gustavson, G. Bonnet, A. Ynnerman, J. Unger, A unified framework for multi-sensor HDR video reconstruction, Signal Process. Image Commun. 29 (2) (2014) 203–215.
[14] J. Froehlich, S. Grandinetti, B. Eberhardt, S. Walter, A. Schilling, H. Brendel, Creating cinematic wide gamut HDR-video for the evaluation of tone mapping operators and HDR-displays, in: Proceedings of SPIE Electronic Imaging, 2014.
[15] S. Nayar, T. Mitsunaga, High dynamic range imaging: spatially varying pixel exposures, in: CVPR, 2000.

[16] S. Hajisharif, J. Kronander, J. Unger, HDR reconstruction for alternating gain (ISO) sensor readout, in: M.W. Eric Galin (Ed.), Eurographics 2014 Short Papers, 2014.
[17] A. Artusi, F. Banterle, T.O. Aydın, D. Panozzo, S.-H. Olga, Image Content Retargeting: Maintaining Color, Tone, and Spatial Consistency, first ed., AK Peters/CRC Press, Natick, MA, 2016, 260 pp.
[18] L. Wang, L.-Y. Wei, K. Zhou, B. Guo, H.-Y. Shum, High dynamic range image hallucination, in: Proceedings of Eurographics Symposium on Rendering, 2007.
[19] C. Ancuti, T. Haber, T. Mertens, P. Bekaert, Video enhancement using reference photographs, Vis. Comput. 24 (7–9) (2008) 709–717.
[20] D.G. Lowe, Distinctive image features from scale-invariant keypoints, Int. J. Comput. Vis. 60 (2) (2004) 91–110.
[21] A. Gupta, P. Bhat, M. Dontcheva, B. Curless, O. Deussen, M. Cohen, Enhancing and experiencing spacetime resolution with videos and stills, in: International Conference on Computational Photography, IEEE, 2009.
[22] N. Snavely, S.M. Seitz, R. Szeliski, Photo tourism: exploring photo collections in 3D, ACM Trans. Graph. 25 (3) (2006) 835–846.
[23] C.L. Zitnick, S.B. Kang, Stereo for image-based rendering using image over-segmentation, Int. J. Comput. Vis. 75 (1) (2007) 49–65.
[24] S.E. Chen, L. Williams, View interpolation for image synthesis, in: Proceedings of the 20th Annual Conference on Computer Graphics and Interactive Techniques (SIGGRAPH'93), ACM, New York, NY, USA, 1993, pp. 279–288.
[25] L. McMillan, G. Bishop, Plenoptic modeling: an image-based rendering system, in: Proceedings of the 22nd Annual Conference on Computer Graphics and Interactive Techniques (SIGGRAPH'95), ACM, New York, NY, USA, 1995, pp. 39–46.
[26] V. Kwatra, A. Schödl, I. Essa, G. Turk, A. Bobick, Graphcut textures: image and video synthesis using graph cuts, in: ACM SIGGRAPH 2003 Papers (SIGGRAPH'03), ACM, New York, NY, USA, 2003, pp. 277–286.
[27] Y. Boykov, O. Veksler, R. Zabih, Fast approximate energy minimization via graph cuts, IEEE Trans. Pattern Anal. Mach. Intell. 23 (11) (2001) 1222–1239.
[28] E. Reinhard, M. Ashikhmin, B. Gooch, P. Shirley, Color transfer between images, IEEE Comput. Graph. Appl. 21 (5) (2001) 34–41.
[29] W.H. Press, S.A. Teukolsky, W.T. Vetterling, B.P. Flannery, Numerical Recipes: The Art of Scientific Computing, third ed., Cambridge University Press, New York, NY, USA, 2007.
[30] P.J. Burt, E.H. Adelson, Readings in computer vision: issues, problems, principles, and paradigms, in: The Laplacian Pyramid as a Compact Image Code, Morgan Kaufmann Publishers Inc., San Francisco, CA, USA, 1987, pp. 671–679.
[31] R. Fattal, D. Lischinski, M. Werman, Gradient domain high dynamic range compression, ACM Trans. Graph. 21 (3) (2002) 249–256.
[32] P. Pérez, M. Gangnet, A. Blake, Poisson image editing, ACM Trans. Graph. 22 (2003) 313–318.
[33] P. Bhat, C.L. Zitnick, M. Cohen, B. Curless, GradientShop: a gradient-domain optimization framework for image and video filtering, ACM Trans. Graph. 29 (2010) 10:1–10:14.
[34] J. Chen, S. Paris, F. Durand, Real-time edge-aware image processing with the bilateral grid, ACM Trans. Graph. 26 (3) (2007) 103.
[35] F. Banterle, M. Corsini, P. Cignoni, R. Scopigno, A low-memory, straightforward and fast bilateral filter through subsampling in spatial domain, Comput. Graph. Forum 31 (1) (2012) 19–32.

CHAPTER 3

HDR Multiview Image Sequence Generation: Toward 3D HDR Video

R.R. Orozco*, C. Loscos†, I. Martin*, A. Artusi*
*University of Girona, Girona, Spain
†University of Reims Champagne-Ardennes, Reims, France

Contents

1. INTRODUCTION

High dynamic range (HDR) content generation has been recently moving from the 2D to 3D imaging domain, introducing a series of open problems that need to be solved. Three-dimensional images are displayed in two main ways: either from two views for monoscopic displays with glasses, or from multiple views for auto-stereoscopic displays. Most current auto-stereoscopic displays accept from five to nine different views [1]. To our knowledge, HDR auto-stereoscopic displays do not exist yet. However, HDR images are device independent, they store values from the scene they represent independently of the device that will project them. Actually, HDR images existed long before the first HDR prototype appeared. Similarly to displaying tone-mapped HDR images on low dynamic range (LDR) displays, it is possible to feed LDR auto-stereoscopic displays with tone-mapped HDRs, one per each view required by the display.

High Dynamic Range Video
http://dx.doi.org/10.1016/B978-0-12-809477-8.00003-0

Some of the techniques used to acquire HDR images from multiple LDR exposures have been recently extended for multiscopic images [2–10]. However, most of these solutions suffer from a common limitation: they rely on accurate dense stereo matching between images that is not robust in case of brightness difference between exposures [11].

This chapter presents a solution to combine sets of multiscopic LDR images into HDR content using image correspondences based on the Patch-Match algorithm [12]. This algorithm was recently used by Sen et al. [13] to build HDR images preventing from significant ghosting effects. Furukawa and Ponce [14] noticed the importance of improving the coherence of neighboring patches, an issue tackled in this chapter. Their results were promising for multiexposure sequences where the reference image is moderately underexposed or saturated, but it fails when the reference image has large underexposed or saturated areas.

The method described in this chapter improves [12] the approach for multiscopic image sequences (Fig. 1). It also reduces the search space in the matching process and improves the incoherence of the matches in the original algorithm. Each image in the set of multiexposed images is used as a reference; we look for matches in all the remaining images. Accurate matches allow us to synthesize a set of HDR images, one for each view used for HDR merging.

The main contributions of this chapter can be summarized as follows:

- We provide an efficient solution to multiscopic HDR image generation.
- Traditional stereo matching produces several artifacts when directly applied on images with different exposures. We introduce the use of an improved version of patch match to solve these drawbacks.
- Patch-match algorithm was adapted to take advantage of the epipolar geometry reducing its computational costs while improving its matching coherence drawbacks.

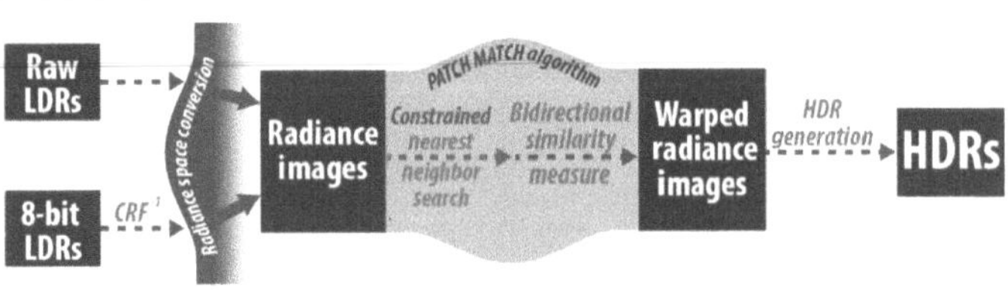

Fig. 1 Proposed framework for multiscopic HDR generation. It is composed of three main steps: (1) radiance space conversion, (2) patch-match correspondences search, and (3) HDR generation.

1.1 Stereoscopic Imaging

Apart from a huge amount of colors and fine details, our visual system is able to perceive depth and 3D shape of objects. Digital images offer a representation of reality projected in 2D arrays. We can guess the distribution of objects in depth because of monoscopic cues like perspective, but we cannot actually perceive depth in 2D images. Our brain needs to receive two slightly different projections of the scene to actually perceive depth.

Stereoscopy is any imaging technique which enhances or enables depth perception using the binocular vision cues [15]. Stereo images refer to a pair of images horizontally aligned and separated at a scalable distance similar to the average distance between human eyes. The different available stereo display systems project them in a way such that each eye perceives only one of the images. In recent years, technologies like stereoscopic cameras and displays have become available to consumers [16–18]. Stereo images require recording at minimum, two views of a scene, one for each eye. However, depending on the display technology, it could be more. Some auto-stereoscopic displays render more than nine different views for an optimal viewing experience [1].

Some prototypes were proposed to acquire stereo HDR content from two or more differently exposed views. Most approaches [2–5, 8, 19, 20] are based on a rig of two cameras placed like a conventional stereo configuration that captures different exposed images. The next sections offer a background of the geometry of stereo systems (Section 1.2) as well as a survey on the different existing approaches for multiscopic HDR acquisition (Section 2).

1.2 Epipolar Geometry

One of the most popular topics of research in computer vision is stereo matching, which refers to the correspondence between pixels of stereo images. The geometry that relates 3D objects to their 2D projection in stereo vision is known as *epipolar geometry*. It explains how the stereo images are related and how depth can mathematically be retrieved from a pair of images.

Fig. 2 describes the main components of the epipolar geometry. A point x in the 3D world coordinates is projected onto the left and right images I_L and I_R, respectively. c_L and c_R are the two centers of projection of the cameras, the plane formed by them, and the point x is known as the epipolar plane. x_L and x_R are the projections of x in I_L and I_R, respectively.

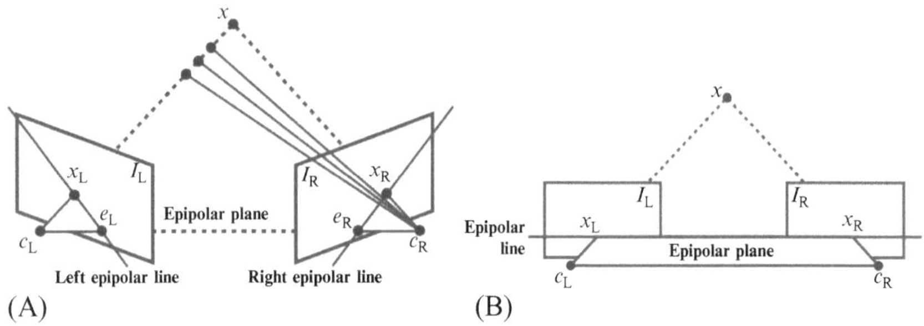

Fig. 2 Main elements of the epipolar geometry. (A) Epipolar geometry. (B) Epipolar geometry rectified.

For any point x_L in the left image, the distance to x is unknown. According to the epipolar geometry, the corresponding point x_R is located somewhere on the right epipolar line. Epipolar geometry does not mean direct correspondence between pixels. However, it reduces the search for a matching pixel to a single epipolar line. The accurate position of a point in the space requires the correct matches between the two images, the focal length, and the distance between the two cameras. Otherwise, only relative measures can be approximated.

If the image planes are aligned and their optical axes are parallel, the two epipolar lines (left and right) converge. In such a case, correspondent pixels rely on the same epipolar line in both images, which simplifies the matching process. Aligning the cameras to force this configuration might be difficult, but images can be aligned.

This alignment process is known as *rectification.* After the images are rectified, the search space for a pixel match is reduced to the same row, which is the epipolar line. To the best of our knowledge, all methods in stereo HDR are based on rectified images and they take advantage of the epipolar constrain during the matching process. Rectified image sets are available on the Internet for testing purposes, like Middlebury [21].

If corresponding pixels (matches) are on the same row on both images, it is possible to define the difference between the images by the horizontal distance between matches in the two images. The image that stores all the horizontal shifts between stereo pairs is called *disparity maps.*

Despite epipolar geometry simplifying the problem, it is far from being solved. Determining pixel matches in regions of similar color is a difficult problem. Moreover, the two views correspond to different projections of the scene, which means that occlusion takes place between objects.

The HDR context adds the fact that different views might be differently exposed, reducing the possibilities of finding color consistent matches.

2. MULTIPLE EXPOSURE STEREO MATCHING

Stereo matching (or disparity estimation) is the process of finding the pixels in the different views that correspond to the same 3D point in the scene. The rectified epipolar geometry simplifies this process to find correspondences on the same epipolar line. It is not necessary to calculate the 3D point coordinates to find the correspondent pixel on the same row of the other image. The disparity is the distance d between a pixel and its horizontal match in the other image.

Akhavan et al. [20, 22] compared the different ways to obtain disparity maps from HDR, LDR, and tone-mapped stereo images. A useful comparison among them is offered and illustrates that the type of input has a significant impact on the quality of the resulting disparity maps.

Fig. 3 shows an example of a differently exposed multiview set corresponding to one frame in a multiscopic system of three views. The main

Fig. 3 LDR multiscopic sequence and the HDR counterpart. *Up*: "Aloe" set of LDR multiview images from Middlebury webpage. *Down*: The tone-mapped HDR result. (A) Multiscopic different exposures. (B) Multiscopic tone-mapped HDR images. *(Images courtesy of Middlebury, Middlebury stereo datasets, http://vision.middlebury.edu/stereo/data/, 2006.)*

goal of stereo matching is to find the correspondences between pixels to generate one HDR image per view for each frame.

Correspondence methods rely on matching cost functions to compute the color similarity between images. It is important to consider that the exposure difference needs to be compensated. Even using radiance space images where pixels are supposed to have the same value for same points in the scene, there might be brightness differences. Such differences may be introduced by the camera due to image noise, slightly different settings, vignetting, or caused by inaccuracies in the estimated CRF. For good analysis and comparison of the existing matching costs and their properties, refer to [9, 11, 23, 24].

Different approaches exist to recover HDR from multiview and multiexposed sets of images. Some of them [3–5] share the same pipeline as in Fig. 4. All mentioned works take as input a set of images with different exposures acquired using a camera with unknown response function. In such cases, the disparity maps need to be calculated in the first instance using the LDR pixel values. Matching images under important differences of brightness is still a big challenge in computer vision.

2.1 Per Frame CRF Recovery Methods

To our knowledge, Troccoli et al. [3] introduced the first technique for HDR recovery from multiscopic images of different exposures. They observed that normalized cross correlation (NCC) is approximately invariant to exposure changes when the camera has a gamma response function. Under such an assumption, they use the algorithm described by Kang and Szeliski [25] to compute the depth maps that maximizes the correspondence between 1 pixel and its projection in the other image. The original approach [25] used sum of squared differences (SSD), but it was substituted by NCC in this work.

Images are warped to the same viewpoint using the depth map. Once pixels are aligned, the CRF is calculated using the method proposed by Grossberg and Nayar [26] over a selected set of matches. With the CRF

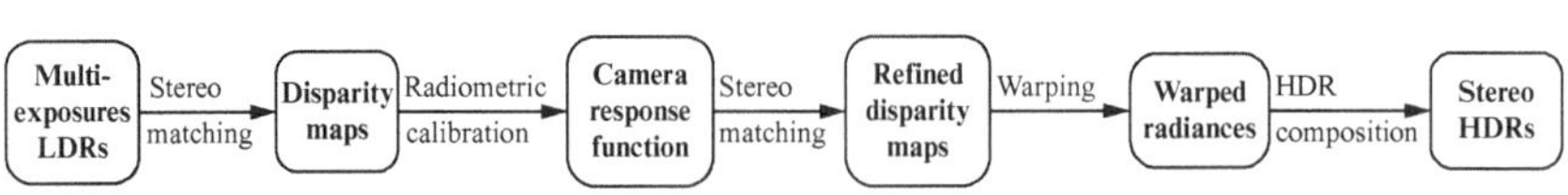

Fig. 4 General multiexposed stereo pipeline for stereo HDR. Proposed by Troccoli et al. [3], used by Sun et al. [5] and Lin and Chang [4], and modified later by Bätz et al. [2].

and the exposure values, all images are transformed to radiance space and the matching process is repeated, this time using SSD. The new depth map improves the previous one and helps to correct artifacts. The warping is updated and HDR values are calculated using a weighted average function.

The same problem was addressed by Lin and Chang [4]. Instead of NCC, they use scale-invariant feature transform (SIFT) descriptors to find matches between LDR stereo images. SIFT is not robust under different exposure images. Only the matches that are coherent with the epipolar and exposure constraints are selected for the next step. The selected pixels are used to calculate the CRF.

The stereo matching algorithm they propose is based on a previous work [27]. Belief propagation is used to calculate the disparity maps. The stereo HDR images are calculated by means of a weighted average function. Even using the best results only, SIFT is not robust enough under significant exposure variations.

A ghost removal technique is used afterward to tackle the artifacts, due to noise or stereo mismatches. The HDR image is exposed to the best exposure of the sequence. The difference between them is calculated and pixels over a threshold are rejected considering them like mismatches. This is risky because HDR values in areas under- and overexposed in the best exposure may be rejected. In this case ghosting would be solved, but LDR values may be introduced in the resulting HDR image.

Sun et al. [5] (inspired by Troccoli et al. [3]) follow the pipeline described in Fig. 4 too. They assume that the disparity map between two rectified stereo images can be modeled as a Markov random field. The matching problem is presented like a Bayesian labeling problem. The optimal disparity values are obtained by minimizing an energy function.

The energy function they use is composed of a pixel dissimilarity term (NCC in their solution) and a disparity smoothness term. It is minimized using the graph cut algorithm to produce initial disparities. The best disparities are selected to calculate the CRF with the algorithm proposed by Mitsunaga and Nayar [28].

Images are converted to radiance space and then another energy minimization is executed to remove artifacts. This time the pixel dissimilarity cost is computed using the Hamming distance between candidates.

The methods presented up to this point have a high computational cost. Calculating the CRF from nonaligned images may introduce errors because the matching between them may not be robust. Two exposures are not enough to obtain a robust CRF with existing techniques. Some of them execute two passes of the stereo matching algorithm, the first one to detect matches for the CRF recovery and a second one to refine the matching results. This might be avoided by calculating the CRF in a previous step using multiple exposures of static scenes. Any of the available techniques [26, 28–30] can be used to get the CRF corresponding to each camera. The curves help to transform pixel values into radiance for each image and the matching process is executed in radiance space images. This avoids one stereo matching step and prevents errors introduced by disparity estimation and image warping.

Rufenacht [19] proposes two different ways to obtain stereoscopic HDR video content. The first, a temporal approach, where exposures are captured by temporally changing the exposure time of two synchronized cameras to get two frames of the same exposure per shot. And the second, a spatial approach, where cameras have different exposure times for all the shots so the two frames of the same shot are exposed differently.

2.2 Offline CRF Recovery Methods

Bonnard et al. [7] propose a methodology to create content that combines depth and HDR video for auto-stereoscopic displays. Instead of varying the exposure times, they use neutral density filters to capture different exposures. A camera with eight synchronized objectives and three pairs of 0.3, 0.6, and 0.9 filters, plus two nonfiltered views provide eight views with four different exposures of the scene stored in 10-bit RAW files. They use a geometry-based approach to recover depth information from epipolar geometry. Depth maps drive the pixel match procedure.

Bätz et al. [2] present a work flow for disparity estimation, divided in the following steps:

- *Cost initialization* consists of evaluating the cost function, Zero NCC in this case, for all values within a disparity search range.

  The matching is performed on the luminance channel of radiance space image using patches of 9×9 pixels. The result of searching for disparities is the disparity space image (DSI), a matrix of $m \times n \times d + 1$ for an images of $m \times n$ pixels with $d + 1$ being the disparity search range.
- *Cost aggregation* smoothes the DSI and finds the final disparity of each pixel in the image. They use an improved version of the cross-based aggregation method described by Mei et al. [31]. This step is performed, not in the luminance channel like in the previous step, but in the actual RGB images.
- *Image warping* is in charge of actually shifting all pixels according to their disparities. Dealing with occluded areas between the images is the main

challenge in this step. The authors propose to do the warping in the original LDR images which adds a new challenge: dealing with under- and overexposed areas. A backward image warping is chosen to implicitly ignore the saturation problems. The algorithm produces a new warped image with the appearance of the reference one by using the target image and the corresponding disparity map. Bilinear interpolation is used to retrieve values at subpixel precision.

Selmanovic et al. [10] propose to generate stereo HDR video from a pair of HDR and LDR videos, using an HDR camera [32] and a traditional digital camera (Canon 1Ds Mark II) in stereo configuration. This chapter is an extension to video of a previous one [15] focused only on stereo HDR images. In this case, one HDR view needs to be reconstructed from two different sources.

Their method proposes three different approaches to generate the HDR:

1. *Stereo correspondence*: It is computed to recover the disparity map between the HDR and the LDR images. The disparity map allows us to transfer the HDR values to the LDR image. The sum of absolute differences (SAD) is used as a matching cost function. Both images are transformed to lab color space, which is perceptually more accurate than RGB.

   The selection of the best disparity value for each pixel is based on *winner takes all* technique. The lower SAD value is selected in each case. An image warping step based on Fehn [33] is used to generate a new HDR image corresponding to the LDR view. The SAD stereo matcher can be implemented to run in real time, but the resulting disparity maps could be noisy and not accurate. The over- and underexposed pixels may end up in a wrong position. In large areas of the same color and hence same SAD cost, the disparity will be constant. Occlusions, reflective or specular objects may cause some artifacts.
2. *Expansion operator*: It could be used to produce an HDR image from the LDR view. Detailed state-of-the-art reports on LDR expansion were previously published [34, 35]. However, in this case, we need the expanded HDR to remain coherent with the original LDR. Inverse tone mappers are not suitable because the resulting HDR image may be different from the acquired one, producing results not possible to fuse through a common binocular vision.

   They propose an expansion operator based on a mapping between the HDR and the LDR image, using the first one as reference. A reconstruction function maps LDR to HDR values (Eq. 1) based on an HDR histogram with 256 bins, putting the same number of HDR values in each bin, as there are in the LDR histogram.

$$RF = \frac{1}{Card(\Omega_c)} \sum_{i=M(c)}^{M(c)+Card(\Omega_c)} C_{\text{hdr}}(i) \tag{1}$$

In Eq. (1), $\Omega_c = \{j = i \ldots N : q_{\text{ldr}}(j) = c\}, c = 0.255$ is the index if a bin Ω_c, $Card(\cdot)$ returns the number of elements in the bin, N is the number of pixels in the image, $q_{\text{ldr}}(j)$ are the intensity values for the pixel j, $M(c) = \sum_0^c Card(\Omega_c)$ is the number of pixels in the previous bin, and q_{hdr} are the intensities of all HDR pixels sorted ascending. RF is used to calculate the look-up table and afterward expansion can be performed directly assigning the corresponding HDR value to each LDR pixel.

The expansion runs in real time, is not view dependent, and avoids stereo matching. The main limitation is again on saturated regions.

3. *Hybrid method* combines the two previous ones. Two HDR images are generated using the previous approaches (stereo matching and expansion operator). Pixels in well-exposed regions are expanded using the first method (expansion operator), while matches for pixels in under- or overexposed regions are found using SAD stereo matching adding a correction step. A mask of under- and oversaturated regions is created, using a threshold for pixels over 250 or below 5. The areas out of the mask are filled in with the expansion operator while the under- or overexposed regions are filled in with an adapted version of the SAD stereo matching to recover more accurate values in over- or underexposed regions.

Instead of having the same disparity over the whole under- or overexposed region, this variant interpolates disparities from well-exposed edges. Edges are detected using a fast morphological edge detection technique described by Lee et al. [36]. Even though some small artifacts may still be produced by the SAD stereo matching in such areas.

Orozco et al. [37] presented a method to generate multiscopic HDR images from LDR multiexposure images. They adapted a patch-match approach [13] to find matches between stereo images using epipolar geometry constrains. This method reduces the search space in the matching process, and includes an improvement of the incoherence problem described for the patch-match algorithm. Each image in the set of multiexposed images is used as a reference, looking for matches in all the remaining images.

These accurate matches allow us to synthesize images corresponding to each view, which are merged into one HDR per view that can be used in auto-stereoscopic displays.

3. PATCH-BASED MULTISCOPIC HDR GENERATION

The input for multiscopic HDR is a sequence of LDR images (formed of RAW or 8-bit RGB data) as shown in the first row of Fig. 5. Each image is acquired from a different viewpoint, usually from a rig of cameras in a stereo distribution, or multiview cameras (see Fig. 6). If the input images are in a 8-bit format, an inverse CRF needs to be recovered for each camera involved in the acquisition. This calibration step is performed only once,

(A) (B) (C)

Fig. 5 Set of LDR multiview images from the IIS Jumble dataset. The *top row* shows five views with different exposure values. The *bottom row* shows HDR images obtained without alignment (A), using Bätz's method (B), and using our proposed patch-match method (C). (A) Nonaligned. (B) Bätz et al. [2]. (C) Our result. *(Courtesy of M. Bätz, T. Richter, J.-U. Garbas, A. Papst, J. Seiler, A. Kaup, High dynamic range video reconstruction from a stereo camera setup, Signal Process. Image Commun. 29 (2) (2014) 191–202.)*

Fig. 6 The Octocam, multiview camera prototype.

using a static set of images for each camera. The inverse of the CRFs is used to transform the input into radiance space. The remaining steps are performed using radiance space values instead of RGB pixels.

An overview of our framework is shown in Fig. 1. The first step is to recover the correspondences between the **n** images of the set. We propose to use a nearest neighbor search algorithm (see Section 3.1) instead of a traditional stereo matching approach. Each image acts like a reference for the matching process. The output of this step is $\mathbf{n} - 1$ warped images for each exposure. Afterward, the warped images are combined into an output HDR image for each view (see Section 3.2).

3.1 Nearest Neighbor Search

For a pair of images I_r and I_s, we compute a nearest neighbor field (NNF) from I_r to I_s using an improved version of the method presented by Barnes et al. [12]. NNF is defined over patches around every pixel coordinate in image I_r for a cost function $\mathbf{D}$ between two patches of images I_r and I_s. Given a patch coordinate $\mathbf{r} \in I_r$ and its corresponding nearest neighbor $\mathbf{s} \in I_s$, $NNF(\mathbf{r}) = \mathbf{s}$. The values of NNF for all coordinates are stored in an array with the same dimensions as I_r.

We start initializing the NNFs using random transformation values within a maximal disparity range on the same epipolar line. Consequently the NNF is improved by minimizing $\mathbf{D}$ until convergence or a maximum number of iterations is reached. Two candidate sets are used in the search phase as suggested by Barnes et al. [12]:

1. *Propagation* uses the known adjacent nearest neighbor patches to improve NNF. It quickly converges, but it may fall in a local minimum.
2. *Random search* introduces a second set of random candidates that are used to avoid local minimums. For each patch centered in pixel v_0, the candidates u_i are sampled at an exponentially decreasing distance v_i previously defined by Barnes et al.:

$$u_i = v_0 + w\alpha^i R_i \tag{2}$$

where R_i is a uniform random value in the interval $[-1, 1]$, w is the maximum value for disparity search, and α is a fixed ratio (1/2 is suggested).

Taking advantage of the epipolar geometry, both search accuracy and computational performance are improved. Geometrically calibrated images allow us to reduce the search space from 2D to 1D domain, consequently reducing the search domain. The random search of matches only operates in the range of maximum disparity in the same epipolar line (1D domain), avoiding a search in 2D space. This significantly reduces the number of samples to find a valid match.

However, the original NNFs approach [12] used in the patch-match algorithm has two main disadvantages, the lack of completeness and coherency. These problems are illustrated in Fig. 7 and the produced artifact in Fig. 8. The lack of coherency refers to the fact that two neighbor pixels in the reference image may match two separated pixels in the source image, like in Fig. 7A. Completeness issues refer to more than 1 pixel in the reference image matching the same correspondence in the source image, as shown in Fig. 7B.

To overcome this drawback, we propose a new distance cost function D by incorporating a coherence term to penalize matches that are not coherent with the transformation of their neighbors. Both Barnes et al. [12] and Sen et al. [13] use the SSD described in Eq. (4) where **T** represents the transformation between patches of **N** pixels in images I_r and I_s. We propose to penalize matches with transformations that differ significantly from it

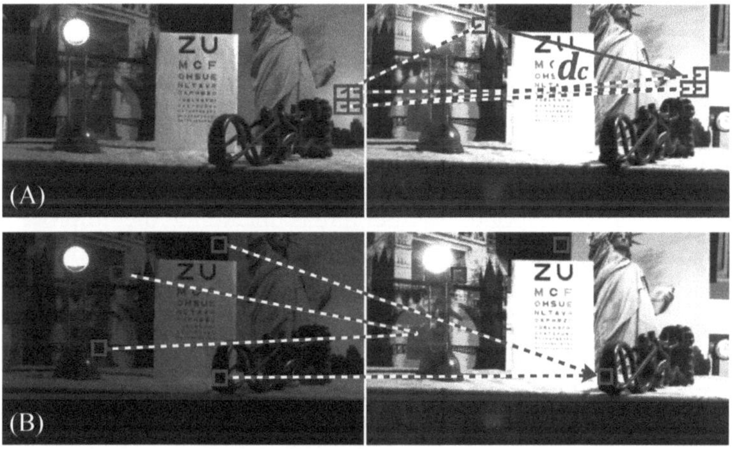

Fig. 7 Patches from the reference image (*left*) look for their NN in the source image (*right*). Even when destination patches are similar in terms of color, matches may be wrong because of geometric coherency problems. (A) Coherency. (B) Completeness. *(Images from the "Octocam" dataset courtesy of J. Bonnard, C. Loscos, G. Valette, J.-M. Nourrit, L. Lucas, High-dynamic range video acquisition with a multiview camera, in: Optics, Photonics, and Digital Technologies for Multimedia Applications II, 2012, pp. 84360A–84360A-11.)*

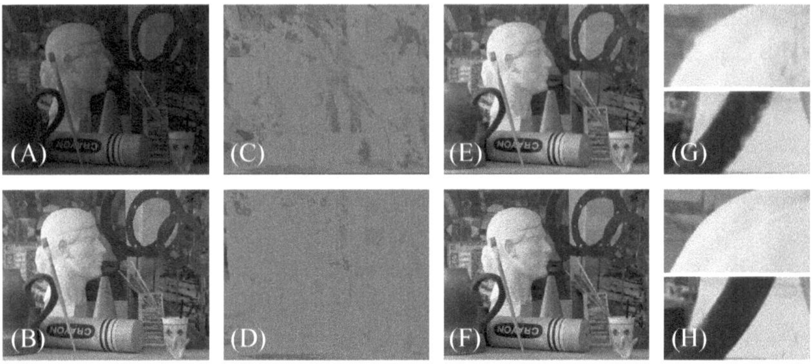

Fig. 8 Matching results using original Patch Match [12] (*Up*) and our version (*Down*) for two iterations using 7 × 7 patches. (A) Src image. (B) Ref image. (C) PM NNF. (D) Ours NNF. (E) PM synthesized. (F) Ours synthesized. (G) Details in (E). (H) Details in (F). *(Images in the "Art" dataset courtesy of Middlebury, Middlebury stereo datasets, http://vision.middlebury.edu/stereo/data/, 2006.)*

neighbors by adding the coherence term **C** defined in Eq. (5). The variable $\mathbf{d_c}$ represents the Euclidean distance to the closest neighbor's match and $\mathbf{Max_{disp}}$ is the maximum disparity value. This new cost function forces pixels to preserve coherent transformations with their neighbors.

$$D = SSD(r,s)/C(r,s) \tag{3}$$

$$SSD = \sum_{n=1}^{N} (I_r - T(I_s))^2 \tag{4}$$

$$C(r,s) = 1 - d_c(r,s)/Max_{disp} \tag{5}$$

Fig. 8D and F corresponds to the results including the improvements presented in this section. Fig. 8C and D shows a color mentioned of the NNFs using HSV color space. The magnitude of the transformation vector is visualized in the saturation channel and the angle in the hue channel. Areas represented with the same color in the NNF color mentioned mean similar transformation. Objects in the same depth may have similar transformation. Notice that the original Patch Match [12] finds very different transformations for neighboring pixels of the same objects and produces artifacts in the synthesized image.

3.2 Image Alignment and HDR Generation

The nearest neighbor search step finds correspondences among all the different views. The matches are stored in a set of $n^2 - n$ NNFs. This information allows to generate $n - 1$ images with different exposures realigned on each view. The set of aligned multiple exposures per view feeds the HDR generation algorithm to produce an HDR image for every view (see Fig. 9).

Despite the improvements in the cost function presented in the previous section, NNF may not be coherent in occluded or saturated areas. However, even in such cases, a match to a similar color is found between each pair of images $I_r; I_s$. This makes it possible to synthesize images for each exposure corresponding to each view.

Direct warping from the NNFs is an option, but it may generate visible artifacts as shown in Fig. 10. We use Bidirectional Similarity Measure (BDSM) (Eq. 6), proposed by Simakov et al. [38] and used by Barnes et al. [12], which measures similarity between pairs of images. The warped images are generated as an average of the patches that contribute to a certain pixel. It is defined in Eq. (6) for every patch $\mathbf{Q} \subset I_r$ and $\mathbf{P} \subset I_s$, and a number $\mathbf{N}$ of patches in each image, respectively. It consists of two terms: *coherence* that ensures that the output is geometrically coherent with the

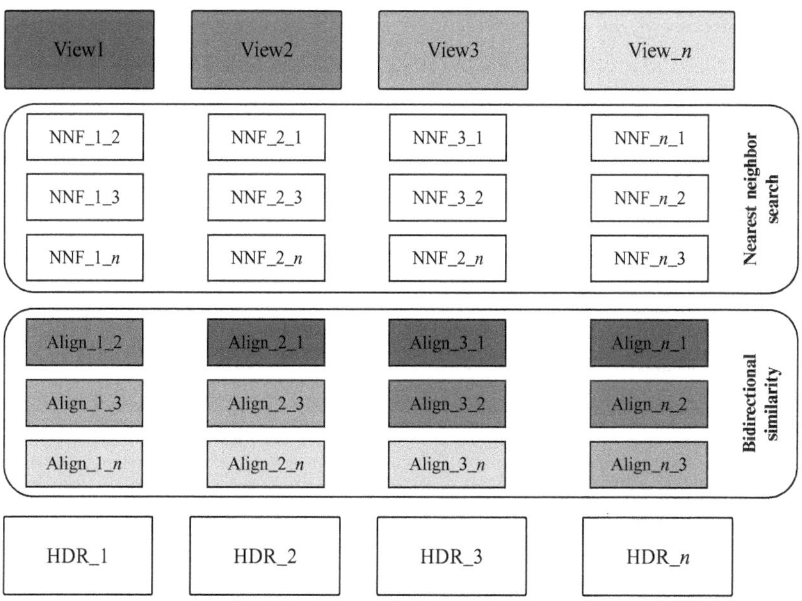

Fig. 9 The nearest neighbor search step generates $n^2 - n$ NNFs. This is used to generate $n - 1$ aligned images per view, using bidirectional similarity. HDR images are generated using input from each view and the corresponding aligned images.

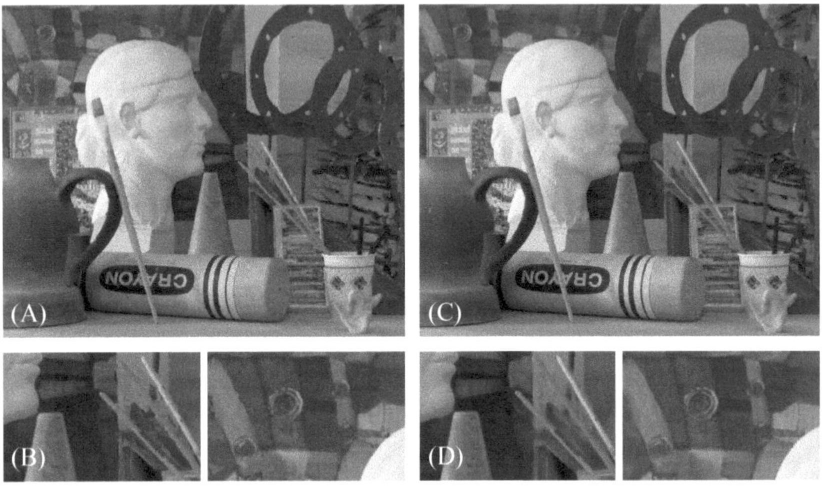

Fig. 10 Images A and C are both synthesized from the pair in Fig. 8. Image A was directly warped using values only from the NNF of Fig. 8C, which corresponds to matching Fig. 8A to B. Image C was warped using the BDSM of Eq. (6) which implies both NNFs of Fig. 8C and D. Notice the artifacts on the edges and the sharp changes on (A) and (B). (A) Direct warping. (B) Details in (A). (C) Using BDSM. (D) Details in (C).

reference and *completeness* that ensures that the output image maximizes the amount of information from the source image:

$$d(I_r, I_s) = \overbrace{\frac{1}{N_{I_r}} \sum_{Q \subset I_r} \min_{P \subset I_s} D(Q, P)}^{d_{\text{completeness}}} + \overbrace{\frac{1}{N_{I_s}} \sum_{P \subset I_s} \min_{Q \subset I_r} D(P, Q)}^{d_{\text{coherence}}} \tag{6}$$

This improves the results by using bidirectional NNFs ($I_r \rightarrow I_s$ and backward, $I_r \leftarrow I_s$). It is more accurate to generate images using only two iterations of nearest neighbor search and bidirectional similarity, than four iterations of neighbor search and direct warping. Table 1 shows some values of mean squared error (MSE) and peak signal-to-noise ratio (PSNR) of images warped like the ones in Fig. 10 comparing to the reference LDR image. The values in the table corresponds to the average MSE and PSNR calculated per each channel of the images in L*a*b* color space, using Eqs. (7) and (8), respectively.

$$MSE(I, I') = \frac{1}{N} \sum_{1=0}^{N} \left(I(i) - I'(i)\right)^2 \tag{7}$$

Table 1 Direct vs bidirectional warping

	Direct warp		Bidirectional	
Iterations	**MSE**	**PSNR**	**MSE**	**PSNR**
1	2.42199	41.8613	2.17003	42.3734
2	2.40966	41.8764	2.17195	42.3762
4	2.4137	41.8728	2.16708	42.3846

$$PSNR(I, I') = 10 \log_{10} \left(\frac{MAX^2}{MSE(I, I')} \right) \tag{8}$$

Because the matching is totally independent for pairs of images, it was implemented in parallel. Each image matches the remaining other views. This produces **n** − 1 NNFs for each view. The NNFs are, in fact, the two components of the BDSM of Eq. (6). The new image is the result of accumulating pixel colors of each overlapping neighbor patch and averaging them.

$$E(i,j) = \frac{\sum_{n=1}^{N} w(I_n(i,j)) \left(\frac{f^{-1}(I_n(i,j))}{\Delta t_n} \right)}{\sum_{n=1}^{N} w(I_n(i,j))} \tag{9}$$

$$w(I_n) = 1 - \left(2\frac{I_n}{255} - 1 \right)^{12} \tag{10}$$

The HDR images (one HDR per view) are generated using a standard weighted average [28–30] as defined in Eq. (9) and the weighting function of Eq. (10) proposed by Khan et al. [39] where I_n represents each image in the sequence, w corresponds to the weight, f is the CRF, Δt_n is the exposure time for the Ith image of the sequence.

4. RESULTS AND DISCUSSION

Five datasets were selected in order to demonstrate the robustness of our results. For the set "Octocam" all the objectives capture the scene at the same time and synchronized shutter speed. For the rest of the datasets, the scenes are static. This avoids the ghosting problem due to dynamic objects in the scene. In all figures of this section, we use the different LDR exposures for display purposes only, the actual matching is performed in radiance space.

The "Octocam" dataset are eight RAW images with 10-bit of color depth per channel. They were acquired simultaneously using the Octocam [40] with a resolution of 748 × 422 pixels. The Octocam is a multiview camera prototype composed by eight objectives horizontally disposed. All images are taken at the same shutter speed (40 ms), but we use three pairs of neutral density filters that reduce the exposure dividing by 2, 4, and 8, respectively. The exposure times for the input sequence are equivalent to 5, 10, 20, and 40 ms, respectively [7]. The objectives are synchronized so all images correspond to the same time instant.

The sets "Aloe," "Art," and "Dwarves" are from the Middlebury website [21]. We selected images that were acquired under fixed illumination conditions with shutter speed values of 125, 500, and 2000 ms for "Aloe" and "Art" and values of 250, 1000, and 4000 ms for "Dwarves." They have a resolution of 1390 × 1110 pixels and were taken from three different views. Even if we have only three different exposures, we can use the seven available views by alternating the exposures as shown in Fig. 14.

The last two datasets were acquired from two of the state-of-the-art papers. Bätz et al. [2] shared their image dataset (IIS Jumble) at a resolution of 2560 × 1920 pixels. We selected five different views from their images. They were acquired at shutter speeds of 5, 30, 61, 122, and 280 ms, respectively. Pairs of HDR images like the one in Fig. 11, both acquired from a scene and synthetic examples, come from Selmanovic et al. [10]. For 8-bit LDR datasets, the CRF is recovered using a set of multiple exposure of a static scene. All LDR images are also transformed to radiance space for fair comparison with other algorithms.

Fig. 11 shows a pair of images linearized from HDR images courtesy of Selmanovic et al. [10] and the comparison between the original PM from Barnes et al. [12] and our method, including the coherence term and epipolar constrains. The images in Fig. 11B and F represent the NNF. They are encoded into an image in HSV color space. Magnitude of the transformation vector is visualized in the saturation channel and the angle in the hue channel. Notice that our results represent more homogeneous transformations, represented in gray color. Images in Fig. 11C and G are synthesized result images for the **Ref** image obtained using pixels only from the **Src** image. The results correspond to the same number of iterations (two in this case). Our implementation converges faster producing accurate results in less iterations than the original method.

All the matching and synthesizing processes are performed in radiance space. They were converted to LDR using the corresponding exposure times, and the CRF for display purposes only. The use of an image synthesis method like the BDSM instead of traditional stereo matching allows us to synthesize values also for occluded areas.

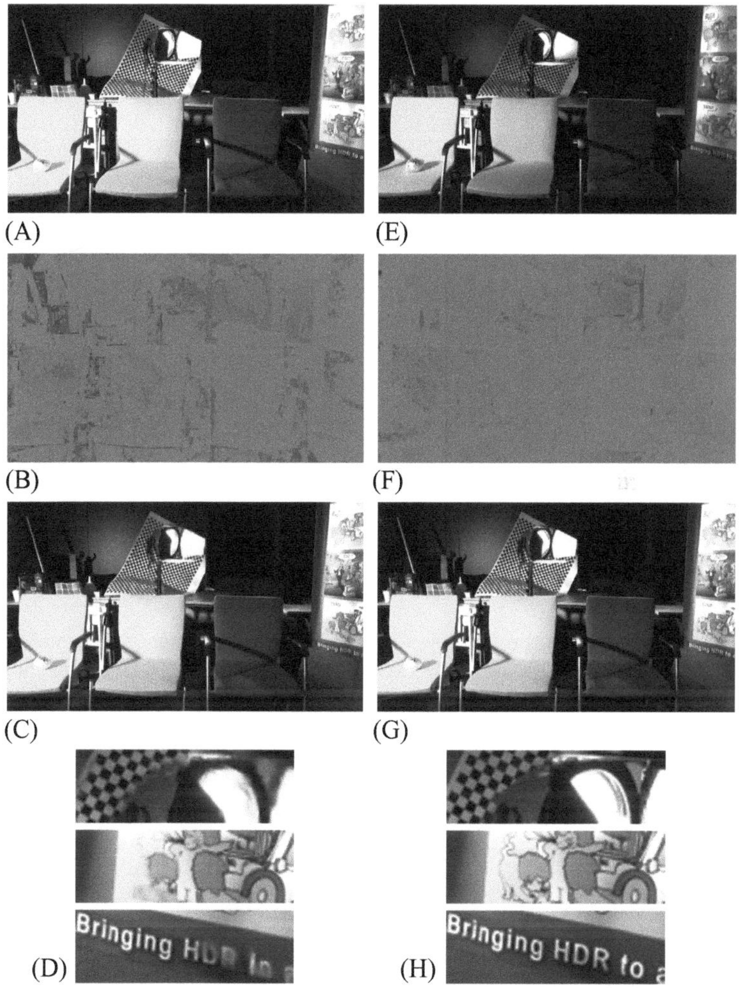

Fig. 11 Comparison between original Patch Match and our method (2 iterations, 7 × 7 patch size). Images B and F show the improvement on the coherence of the NNF, using our method. (A) Src image. (B) PM NNF. (C) PM synthesized. (D) Details in (E). (E) Ref image. (F) Ours NNF. (G) Ours synthesized. (H) Details in (F). *(Images courtesy of E. Selmanovic, K. Debattista, T. Bashford-Rogers, A. Chalmers, Enabling stereoscopic high dynamic range video, Signal Process. Image Commun. 29 (2) (2014) 216–228.)*

Fig. 12 shows the NNFs and the images synthesized for different iterations of both our method and the original patch match. Our method converges faster and produces more coherent results than Barnes et al. [12]. In occluded areas, the matches may not be accurate in terms of geometry

Fig. 12 Two images from the "Dwarves" set of LDR multiview images from Middlebury [21]. Our method (*up row*) with only two iterations achieves very accurate matches. Notice that the original patch (*bottom row*) match requires more iterations to achieve good results in fine details of the image. (A) Reference. (B) Source. (C) One iteration. (D) One iteration. (E) Two iterations. (F) Two iterations. (G) Ten iterations. (H) Ten iterations.

due to the lack of information. Even in such cases, the result is accurate in terms of color. After several tests, only two iterations of our method were enough to get good results while five iterations were recommended for previous approaches.

Fig. 13 shows one example of the generated HDR corresponding to the lowest exposure LDR view in the IIS Jumble dataset. It is the result of merging all synthesized images obtained with the first view as reference. The darker image is also the one that contains more noisy and underexposed areas. HDR values were recovered even for such areas and no visible artifacts appears. On the contrary, the problem of recovering HDR values for saturated areas in the reference image remains unsolved. When the dynamic range differences are extreme, the algorithm does not provide accurate results. Future work must provide new techniques, because the lack of information inside saturated areas does not allow patches to find good matches.

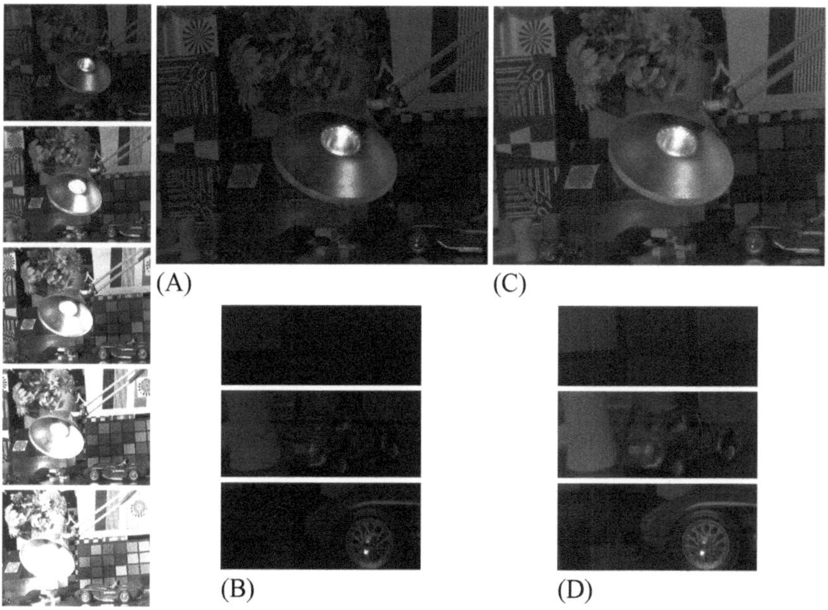

Fig. 13 Details of the generated HDR image corresponding to a dark exposure. Notice that underexposed areas, traditionally difficult to recover, are successfully generated without visible noise or misaligned artifacts. (A) Lower exposure LDR. (B) Details in (B). (C) Tone-mapped HDR. (D) Details in (C). *(IIS Jumble dataset courtesy of M. Bätz, T. Richter, J.-U. Garbas, A. Papst, J. Seiler, A. Kaup, High dynamic range video reconstruction from a stereo camera setup, Signal Process. Image Commun. 29 (2) (2014) 191–202.)*

The inverse CRFs for the LDR images were calculated from a set of aligned multiexposed images using the software RASCAL, provided by Mitsunaga and Nayar [28]. Fig. 14 shows the result of our method for a whole set of LDR multiview and differently exposed images. All obtained images are accurate in terms of contours, no visible artifacts compared to the LDR were obtained.

Fig. 15 shows the result of the proposed method in a scene with important lighting variations. The presence of the light spot introduces extreme lighting differences between the different exposures. For bigger exposures, the light glows from the spot and saturates pixels, not only inside the spot, but also around it. There is not information in saturated areas and the matching algorithm does not find good correspondences. The dynamic range is then compromised in such areas and they remain saturated.

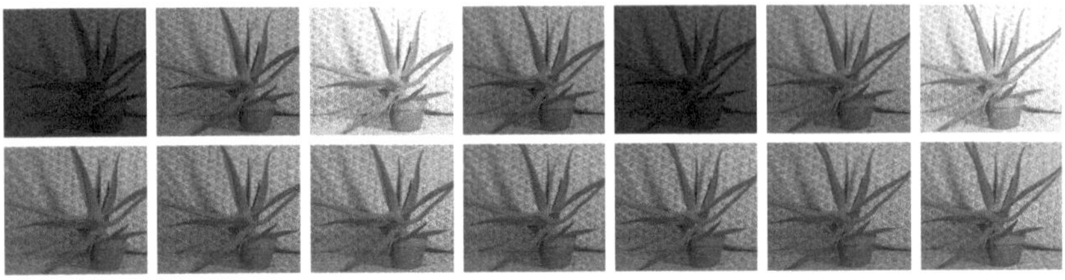

Fig. 14 *Up*: "Aloe" set of LDR multiview images from Middlebury [21]. *Down*: The resulting tone-mapped HDR taking each LDR as reference, respectively. Notice the coherence between the tone-mapped HDR images.

Fig. 15 *Up*: Set of LDR multiview images acquired using the Octocam [40]. *Down*: The resulting tone-mapped HDR, taking each LDR as reference, respectively. Despite the important exposure differences of the LDR sequence, coherent HDR results are obtained. However, highly saturated areas might remain saturated in the resulting HDR. *(Images courtesy of J. Bonnard, G. Valette, J.-M. Nourrit, C. Loscos, Analysis of the consequences of data quality and calibration on 3D HDR image generation, in: European Signal Processing Conference (EUSIPCO), Lisbonne, Portugal, 2014.)*

Two of the datasets used in the tests provide aligned multiple exposures for each view, which allows us to generate ground truth HDR images per view. Fig. 16 shows the results of comparing some of our results to ground truth images using the HDR-VDP-2 metric proposed by Mantiuk et al. [41]. This metric provides some values to describe how similar two HDR images are.

The quality correlate Q is 100 for the best quality and gets lower for lower quality. Q can be negative in case of very large differences. The images at the right of each pairs in Fig. 16 are the probability of detection map. It shows where and how likely a difference will be noticed. However, it does not show what this difference is. Images at the left on each pair show the contrast-normalized per-pixel difference weighted by the probability of detection. The resulting images do not show probabilities. However, they better correspond to the perceived differences.

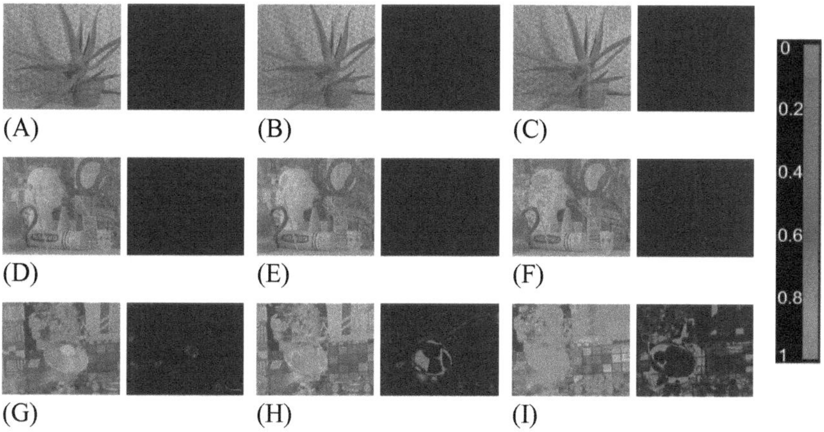

Fig. 16 HDR-VDP-2 comparison between ground truth HDR images and our results. Each pair corresponds to the probability of detection (*left*) and the contrast-normalized per-pixel difference (*right*) of low, medium, and high exposures corresponding to different views. The *first row* corresponds to the three first views of the "Aloe" dataset (Fig. 14). The *second row* to the "Art" dataset (Fig. 8) and the *third row* to the IIS Jumble dataset (Fig. 13) HDR-VDP-2 comparison. (A) $Q=64.7686$. (B) $Q=65.8598$. (C) $Q=62.7004$. (D) $Q=72.3828$. (E) $Q=74.7925$. (F) $Q=65.8817$. (G) $Q=52.3245$. (H) $Q=37.7078$. (I) $Q=29.4814$. *(Images courtesy of Middlebury, Middlebury stereo datasets, http://vision.middlebury.edu/stereo/data/, 2006 and M. Bätz, T. Richter, J.-U. Garbas, A. Papst, J. Seiler, A. Kaup, High dynamic range video reconstruction from a stereo camera setup, Signal Process. Image Commun. 29 (2) (2014) 191–202, respectively.)*

The results illustrate that in general, no differences are perceived. Except in areas that appear totally saturated in the reference image, like the head of the sculpture in the "Art" dataset or the lamp in the IIS Jumble. In such cases visible artifact appears because the matching step fails to find valid correspondences.

Our method is faster than some previous solutions. Sen et al. [13] mention that their method takes less than 3 min for a sequence of seven images of 1350×900 pixels. The combination of a reduced search space and the coherence term effectively implies a reduction of the processing time. On a Intel Core i7-2620M 2.70 GHz with 8 GB of memory, our method takes less than 2 min (103 ± 10 s) for the Aloe dataset with a resolution of 1282×1110 pixels.

REFERENCES

[1] L. Lucas, C. Loscos, Y. Remion, 3D Video From Capture to Diffusion, Wiley-ISTE, New York, NY, 2013.
[2] M. Bätz, T. Richter, J.-U. Garbas, A. Papst, J. Seiler, A. Kaup, High dynamic range video reconstruction from a stereo camera setup, Signal Process. Image Commun. 29 (2) (2014) 191–202.
[3] A. Troccoli, S.B. Kang, S. Seitz, Multi-view multi-exposure stereo, in: Third International Symposium on 3D Data Processing, Visualization, and Transmission, 2006, pp. 861–868.
[4] H.-Y. Lin, W.-Z. Chang, High dynamic range imaging for stereoscopic scene representation, in: 2009 16th IEEE International Conference on Image Processing (ICIP), 2009, pp. 4305–4308.
[5] N. Sun, H. Mansour, R. Ward, HDR image construction from multi-exposed stereo LDR images, in: Proceedings of the IEEE International Conference on Image Processing (ICIP), Hong Kong, 2010.
[6] M. Bleyer, C. Rhemann, C. Rother, PatchMatch stereo-stereo matching with slanted support Windows, in: Proceedings of the British Machine Vision Conference, BMVA Press, 2011, pp. 14.1–14.11, http://dx.doi.org/10.5244/C.25.14.
[7] J. Bonnard, C. Loscos, G. Valette, J.-M. Nourrit, L. Lucas, High-dynamic range video acquisition with a multiview camera, in: Optics, Photonics, and Digital Technologies for Multimedia Applications II, 2012, pp. 84360A–84360A–11.
[8] R.R. Orozco, I. Martin, C. Loscos, A. Artusi, Patch-based registration for auto-stereoscopic HDR content creation, in: HDRi2013—First International Conference and SME Workshop on HDR Imaging, Oporto, Portugal, 2013.
[9] R.R. Orozco, I. Martin, C. Loscos, A. Artusi, Génération de séquences d'images multivues HDR: vers la vidéo HDR, in: 27es journées de l'Association française d'informatique graphique et du chapitre français d'Eurographics, Reims, France, 2014.
[10] E. Selmanovic, K. Debattista, T. Bashford-Rogers, A. Chalmers, Enabling stereoscopic high dynamic range video, Signal Process. Image Commun. 29 (2) (2014) 216–228. Special Issue on Advances in High Dynamic Range Video Research.
[11] J. Bonnard, G. Valette, J.-M. Nourrit, C. Loscos, Analysis of the consequences of data quality and calibration on 3D HDR image generation, in: European Signal Processing Conference (EUSIPCO), Lisbonne, Portugal, 2014.
[12] C. Barnes, E. Shechtman, A. Finkelstein, D.B. Goldman, PatchMatch: a randomized correspondence algorithm for structural image editing, ACM Trans. Graph. (Proc. SIGGRAPH) 28 (3) (2009), http://gfx.cs.princeton.edu/pubs/Barnes_2009_PAR/bibtex.php.
[13] P. Sen, N.K. Kalantari, M. Yaesoubi, S. Darabi, D.B. Goldman, E. Shechtman, Robust patch-based HDR reconstruction of dynamic scenes, ACM Trans. Graph. (Proc. SIGGRAPH Asia 2012) 31 (6) (2012) 203:1–203:11.
[14] Y. Furukawa, J. Ponce, Accurate, dense, and robust multiview stereopsis, IEEE Trans. Pattern Anal. Mach. Intell. 32 (8) (2010) 1362–1376.
[15] E. Selmanović, K. Debattista, T. Bashford-Rogers, A. Chalmers, Generating stereoscopic HDR images using HDR-LDR image pairs, ACM Trans. Appl. Percept. 10 (1) (2013) 3:1–3:18.
[16] H. Urey, K.V. Chellappan, E. Erden, P. Surman, State of the art in stereoscopic and autostereoscopic displays, Proc. IEEE 99 (2011) 540–555.
[17] B. Mendiburu, Y. Pupulin, S. Schklair (Eds.), 3D {TV} and 3D Cinema, Focal Press, Boston, MA, 2012.
[18] F. Dufaux, B. Pesquet-Popescu, M. Cagnazzo, Emerging Technologies for 3D Video: Creation, Coding, Transmission and Rendering, John Wiley & Sons, New Jersey, USA, 2013.

[19] D. Rufenacht, Stereoscopic High Dynamic Range Video, PhD Thesis, Ecole Polytechnique Fédérale de Lausanne (EPFL), Switzerland, 2011.

[20] T. Akhavan, C. Kapeller, J.-H. Cho, M. Gelautz, Stereo HDR disparity map computation using structured light, in: HDRi2014 Second International Conference and SME Workshop on HDR Imaging, 2014.

[21] Middlebury, Middlebury stereo datasets, http://vision.middlebury.edu/stereo/data/, 2006.

[22] T. Akhavan, H. Yoo, M. Gelautz, A framework for HDR stereo matching using multi-exposed images, in: Proceedings of HDRi2013—First International Conference and SME Workshop on HDR Imaging, The Eurographics Association and Blackwell Publishing Ltd., Paper No. 8, Oxford/Malden, 2013.

[23] D. Scharstein, R. Szeliski, A taxonomy and evaluation of dense two-frame stereo correspondence algorithms, Int. J. Comput. Vis. 47 (1) (2002) 7–42.

[24] H. Hirschmuller, D. Scharstein, Evaluation of stereo matching costs on images with radiometric differences, IEEE Trans. Pattern Anal. Mach. Intell. 31 (9) (2009) 1582–1599.

[25] S.B. Kang, R. Szeliski, Extracting view-dependent depth maps from a collection of images, Int. J. Comput. Vis. 58 (2) (2004) 139–163.

[26] M.D. Grossberg, S.K. Nayar, Determining the camera response from images: what is knowable?, IEEE Trans. Pattern Anal. Mach. Intell. 25 (11) (2003) 1455–1467.

[27] J. Sun, N.-N. Zheng, H.-Y. Shum, Stereo matching using belief propagation, IEEE Trans. Pattern Anal. Mach. Intell. 25 (7) (2003) 787–800.

[28] T. Mitsunaga, S.K. Nayar, Radiometric self calibration, IEEE Int. Conf. Comput. Vis. Pattern Recogn. 1 (1999) 374–380.

[29] S. Mann, R.W. Picard, On Being Undigital With Digital Cameras: Extending Dynamic Range by Combining Differently Exposed Pictures, Perceptual Computing Section, Media Laboratory, Massachusetts Institute of Technology, Cambridge, MA, 1995, pp. 442–448.

[30] P. Debevec, J. Malik, Recovering high dynamic range radiance maps from photographs, in: Proceedings of ACM SIGGRAPH (Computer Graphics), vol. 31, ACM Press/Addison-Wesley Publishing Co., New York, NY, 1997, pp. 369–378.

[31] X. Mei, X. Sun, M. Zhou, S. Jiao, H. Wang, X. Zhang, On building an accurate stereo matching system on graphics hardware, in: 2011 IEEE International Conference on Computer Vision Workshops (ICCV Workshops), 2011, pp. 467–474.

[32] A. Chalmers, G. Bonnet, F. Banterle, P. Dubla, K. Debattista, A. Artusi, C. Moir, High-dynamic-range video solution, in: ACM SIGGRAPH ASIA 2009 Art Gallery & #38; Emerging Technologies: Adaptation, SIGGRAPH ASIA '09, ACM, New York, NY, 2009, pp. 71–71, http://dl.acm.org/citation.cfm?id=1665190.

[33] C. Fehn, Depth-image-based rendering (DIBR), compression, and transmission for a new approach on 3D-TV, in: Proc. SPIE, vol. 5291, 2004, pp. 93–104.

[34] F. Banterle, K. Debattista, A. Artusi, S. Pattanaik, K. Myszkowski, P. Ledda, A. Chalmers, High dynamic range imaging and low dynamic range expansion for generating HDR content, Comput. Graph. Forum 28 (8) (2009) 2343–2367.

[35] K. Hirakawa, P.M. Simon, Single-shot high dynamic range imaging with conventional camera hardware, in: 2011 IEEE International Conference on Computer Vision (ICCV), 2011, pp. 1339–1346.

[36] J. Lee, R.M. Haralick, L.G. Shapiro, Morphologic edge detection, IEEE J. Robot. Autom. 3 (2) (1987) 142–156.

[37] R.R. Orozco, I. Martin, C. Loscos, A. Artusi, Multiscopic HDR image sequence generation, J. WSCG 23 (2) (2015) 111–120.

[38] D. Simakov, Y. Caspi, E. Shechtman, M. Irani, Summarizing visual data using bidirectional similarity, in: IEEE Conference on Computer Vision and Pattern Recognition 2008 (CVPR'08), 2008.

[39] E.A. Khan, A.O. Akyüz, E. Reinhard, Ghost removal in high dynamic range images, IEEE Int. Conf. Image Process. (2006) 2005–2008, http://ieeexplore.ieee.org/document/4106952/.
[40] J. Prévoteau, S. Chalençcon-Piotin, D. Debons, L. Lucas, Y. Remion, Multi-view shooting geometry for multiscopic rendering with controlled distortion, Int. J. Dig. Multimedia Broadcast. 2010 (2010) 1–11.
[41] R. Mantiuk, K.J. Kim, A.G. Rempel, W. Heidrich, HDR-VDP-2: a calibrated visual metric for visibility and quality predictions in all luminance conditions, ACM Trans. Graph. 30 (4) (2011) 40:1–40:14.

CHAPTER 4

Calibrated Measurement of Imager Dynamic Range

B. Karr[*,†], K. Debattista[*], A. Chalmers[*]
[*] University of Warwick, Coventry, United Kingdom
[†] Rockledge Design Group, Inc., Rockledge, FL, United States

Contents

1. INTRODUCTION

Traditional imaging methods are unable to capture the wide range of luminance present in a natural scene. High dynamic range (HDR) imaging is an exception, enabling a much larger range of light in a scene to be captured [1, 2], surpassing the simultaneous capabilities of human vision. As interest in HDR grows, especially with its inclusion in the UHDTV definition (ITU-R Recommendation BT.2020), a number of imaging vendors are now beginning to offer capture systems which they claim to be HDR. To validate such claims, a new method is presented that evaluates a no-reference objective quality metric (NROQM) of image system dynamic range across a wider luminance range, utilizing off-the-shelf equipment and software.

Dynamic range (DR) as it relates to noise can be viewed as one aspect of image quality, the evaluation of which can be broadly categorized as either subjective or objective. Subjective evaluation generally utilizes a large

High Dynamic Range Video
http://dx.doi.org/10.1016/B978-0-12-809477-8.00004-2

sample of human observers and a carefully designed experiment, often based on a standardized test procedure [3, 4]. Subjective methods are considered to be the most reliable when considering the human vision system (HVS), however they can be costly, time consuming, and demanding when considering the knowledge and effort required to obtain meaningful results [2]. Objective evaluation has the goal of accurately predicting standard subjective quality ratings, and when selected and designed for a specific application, can also produce reasonable cost-efficient results [5]. Objective evaluations make use of computational methods to assess the quality of images, thus human factors such as mood, past experience, and viewing environment are eliminated. Tests can be practically performed for large test samples, and repeatable results can be obtained utilizing identical test conditions. Objective measurements can be classified as full-reference (FR) when a full reference image is available for comparison, reduced-reference (RR) when a subset of feature characteristics are available, and no-reference (NR) when the focus is made purely on distortions, such as discrete cosine transform (DCT) encoding blockiness, wavelet type encoding blurring, or noise evaluation. Examples of FR methods include the HDR visual quality measure (HDR-VQM) [5] and visual difference predictor (HDR-VDP) [6], which are based on spatiotemporal analysis of an error video where the localized perceptual error between a source and a distorted video is determined. Methods such as HDR-VDP require complex calibration of optical and retinal parameters. Alternatives implementing simple arithmetic on perceptually linearized values include peak signal-to-noise ratio (PSNR) and structural similarity index (SSIM) [7]. RR methods are often used in transmission applications where the full original image cannot, or is not, transmitted with the compressed image data. A reduced set of criteria is selected for RR to produce an objective quality score based on the reduced description (RD) of the reference and the distorted image [8]. As an alternative, no-reference models are typically built from individual or combined metrics, such as sharpness, contrast, clipping, ringing, blocking artifacts, and noise [9]. In the case of our DR calculation, the no-reference metric is fundamentally based on the evaluation of noise, using a single frame of a 21 density step HDR test chart as an input to determine the ratio of the maximum unclipped input luminance to the minimum acceptable input luminance level meeting specified noise criteria. The minimum acceptable input luminance is calculated for several values of signal-to-noise ratio (SNR), with results presented in terms of the relative luminance units of stops of noise. Corresponding arbitrary quality ratings are discussed as they relate to the measurement and perceived quality.

This chapter evaluates the use of a noise as a NROQM to accurately measure and transparently compare the increased DR capability of modern image capture systems, and examines the arbitrary assignment of perceived quality terms to the results. The method described requires a reduced set of test equipment, primarily an HDR test chart and processing software, and can be performed without having to determine the sensor specific response function. A review of image sensor theory, the associated noise model, and calculation theory is presented first. The test methodology is described next, including data workflow and analysis procedures. Results are presented for a number of modern imaging systems, including some capable of HDR capture, along with manufacturer stated capabilities. Finally, the results are discussed as they relate to perceived DR quality and conclusions drawn.

1.1 Background

DR is defined as the ratio of luminance of the lightest and darkest elements of a scene or image. In an absolute sense, this ratio can be considered the brightest to darkest pixel, but more commonly it is defined as the ratio of the largest nonsaturating signal to one standard deviation above the camera noise under dark conditions [10, 11]. An equation for DR_{SNR} can be described as the ratio of maximum to minimum input luminance level:

$$DR_{SNR} = \frac{L_{sat}}{L_{min}}, \tag{1}$$

where L_{sat} is the maximum unclipped input luminance level, defined as 98% of its maximum value (250 of 255 for 24 bit color), and L_{min} is the minimum input luminance level for a specified SNR. The ratio can be specified as instantaneous for a single instant of time, or temporal, where the luminance values change over time, that is, light and dark elements over a series of images. Poynton [12] uses the definitions of simultaneous for the instantaneous case, and sequential for the temporal case. The primary focus of this chapter relates to simultaneous measurement, where a single frame of an HDR test chart from a video sequence is analyzed. New classes of imaging sensors are being developed that are capable of directly capturing larger DR on the scale of 5 orders of magnitude (5 $\log_{10}$, 16.6 stops) [13, 14]. For comparison, human vision generally is agreed to have a simultaneous DR capability of approximately four orders of magnitude (4 $\log_{10}$, 13.3 stops) [1, 15], and a sequential DR via adaption of approximately 10 orders of magnitude (10 $\log_{10}$, 33.2 stops) [16, 17].

In order to objectively calculate the capture system DR, an understanding of the imaging and noise transfer function is required. An image sensor measures the radiant power (flux) Φ at each pixel j for each image i. The amount of photons n_{p_j} that are absorbed at each pixel area A_j over time period t_i is given by:

$$n_{p_j} = \frac{A_j u_j t_i}{E_\mathrm{p}}, \tag{2}$$

where μ_j is the mean incident spectral irradiance at pixel j. E_p is the photon energy defined by:

$$E_\mathrm{p} = \frac{hc}{\lambda}, \tag{3}$$

with the wavelength of the photon given by λ, the speed of light $c = 2.9979 \times 10^8$ m/s, and Planck's constant $h = 6.626 \times 10^{-34}$ Js. Photons incident on the pixel over the exposure time generate an average number of electrons n_p based on the quantum efficiency of the sensor $\eta(\lambda)$:

$$\eta(\lambda) = \frac{n_\mathrm{p}}{n_e}, \tag{4}$$

producing a photocurrent i_ph and photon noise n_p at each pixel site. Thermally dependent leakage current in the electronics (free electrons in the silicon), known as dark current i_d contributes to the total charge, as shown in Fig. 1. In conventional current mode, the photocurrent produces a charge Q_j on a capacitive integrator generating a voltage V_j amplified with gain g. The voltage gain product $V_j g$ includes readout noise n_r and amplifier noise n_a. Typically, a buffer circuit follows the gain amplifier contributing an additional buffer amplifier noise n_b, leading to an analog to digital converter (ADC) with quantization noise n_q, resulting in the measured pixel digital number (DN) d_{ij}. In linear systems, the DN is proportional to the number of electrons collected with an overall system gain G of units of DN/*electrons*.

Given a linear transfer model, the noise sources can be summed and we can write an equation for the total system noise at the output of the ADC as:

$$n_\mathrm{total}(j) = n_\mathrm{p}(j) + n_\mathrm{d}(j) + n_\mathrm{r}(j) + n_\mathrm{a}(j) + n_b(j) + n_q(j). \tag{5}$$

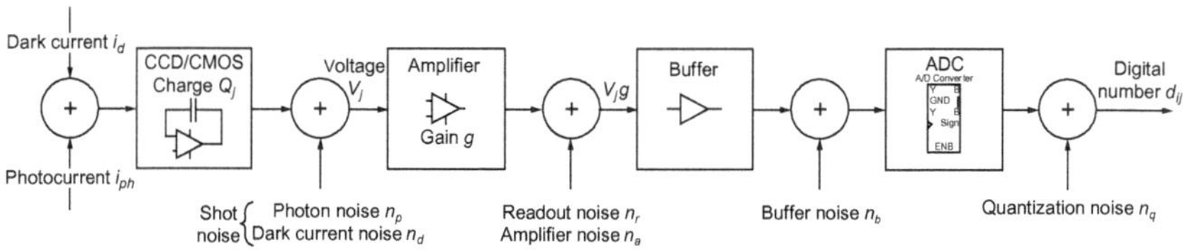

Fig. 1 Image and noise transfer model.

When specifying noise units, it is understood that the noise magnitude is the RMS value of the random process producing the noise [10]. If n_{postgain} is defined as the sum of the postgain amplifier noise sources n_{r}, n_{a}, n_{b}, and n_{q}, the variance is given as:

$$\sigma^2_{n_{\text{total}}} = \sigma^2_{n_{\text{p}}} + \sigma^2_{n_{\text{d}}} + \sigma^2_{n_{\text{postgain}}}. \tag{6}$$

The variance represents the random noise power of the system, and the standard deviation characterizes the system RMS noise [18]. Finally, Reibel et al. [18] note that no system is strictly linear, and introduce a nonlinear contribution C_{NL} to describe the signal variance that is modeled by a dependence on the total electrons transferred to the sense node capacitor (photoelectrons and dark current electrons) and the overall system gain G resulting in:

$$\sigma^2_{n_{\text{total}}} = \sigma^2_{n_{\text{p}}} + \sigma^2_{n_{\text{d}}} + \sigma^2_{n_{\text{postgain}}} + C_{\text{NL}}. \tag{7}$$

A linear transfer model not only permits the summation of noise sources, it also ensures the relative brightness of the scene is maintained. To achieve linearity over the entire sensor, and to minimize fixed pattern noise (including streaking and banding), a radiometric camera calibration is generally required for each pixel, resulting in a total camera response function. The calibrated camera response function can be combined with optical characterization and other camera postprocessing functions such as gamma, tonal response, noise reduction, and sharpening to produce an overall camera opto-electronic conversion function (OECF) [19, 20].

Noise is meaningful in relation to a signal S (i.e., SNR) and can be defined differently depending on how the measurement is performed. If the measurement is concerned with the sensor itself, isolated from the camera postprocessing functions described above, noise can be referenced to the original scene. This can be accomplished by linearizing the data using the inverse OECF, and defining the signal as a pixel difference corresponding to a specified scene density range. Alternatively, if raw data is unavailable or the OECF is unknown (or beyond the scope or measurement capability), or if the focus is on complete system capability (optics, sensor, and camera performance) rather than sensor specific, the signal can be defined with respect to an individual patch pixel level. The evaluation of DR described in this chapter is based on the latter, with a goal of describing a straightforward method that can be used to produce an ever-increasing database of comparable camera system DR measurements. If the variance σ^2_{signal} of the signal is known and the signal is zero mean, the SNR for the minimum input luminance can be expressed as:

$$\text{SNR}_{L_{\min}} = \frac{\sigma^2_{\text{signal}}}{\sigma_{n^2_{\text{total}}}} \tag{8}$$

where $\sigma^2_{n_{\text{total}}}$ is given by Eq. (6). Alternatively, Eq. (8) can be expressed as the ratio of the mean pixel value μ_{mean} to the standard deviation $\sigma_{\text{Std Dev}}$, and equivalently to the total RMS noise $n_{\text{total}_{\text{RMS}}}(j)$ as:

$$\text{SNR}_{L_{\min}} = \frac{\mu_{\text{mean}}}{\sigma_{\text{Std Dev}}} = \frac{\mu_{\text{mean}}(j)}{n_{\text{total}_{\text{RMS}}}(j)}. \tag{9}$$

Minimum input luminance level is often determined from the density step having a SNR of 1 or greater [11, 21]. The processing software utilized in this study named Imatest [22], performs automatic calculations for $\text{SNR}_{L_{\min}}$ values of 1, 2, 4, and 10. It can be convenient to reevaluate Eq. (9) in terms of stops. Stops is a common photography term where one stop is equivalent to a halving or doubling of light, corresponding more closely to the relative luminance response of human vision [23]. We can express $\text{SNR}_{L_{\min}}$ in terms of noise in stops by:

$$n_{\text{stops}_{L_{\min}}} = \frac{1}{\text{SNR}_{L_{\min}}}, \tag{10}$$

where a $\text{SNR}_{L_{\min}}$ of 1 is equivalent to 1 stop of noise. Fig. 2 illustrates the assignment of the arbitrary quality descriptions (high, medium-high, medium, and low) for 0.1, 0.25, 0.5, and 1 stop of noise as calculated and defined in Imatest, as well as the corresponding SNR values in terms of a luminance ratio as well as in decibels (dB). Finally, DR_{SNR} can also be expressed in terms of relative luminance or stops as:

$$\text{DR}_{\text{stops}} = \frac{\log_{10}(L_{\text{sat}}) - \log_{10}\left(L_{\min_{\text{SNR}}}\right)}{\log_{10}(2)}. \tag{11}$$

The question that remains is: How are the arbitrary quality descriptors chosen and why? We can start to answer this question by reviewing early 1940s era work in human signal detection by Albert Rose, known as the Rose model. The Rose model provided a good approximation of a Bayesian ideal observer, albeit for carefully and narrowly defined conditions [24]. Rose defined a constant *k*, to be determined experimentally, as the threshold SNR for reliable detection of a signal. The value of *k* was experimentally estimated by Rose to be in the region of 3–7, based on observations of photographic film and television pictures. Additional experiments using a light spot scanner arrangement resulted in an estimate of *k* of approximately 5, known as the Rose criterion, stating that a *SNR* of at least 5 (13.98 dB) is required to distinguish features at 100% certainty. The early empirical determinations of *k* had several issues. Dependency on a specification of percent chance of detection (i.e., 90%, 50%), included limitations of human vision, such as spatial integration and contrast sensitivity, and included dependence on viewing distance amongst other factors. The development of signal detection theory (SDT) and objective experimental techniques worked to overcome these limitations [25, 26]. For comparison, a binary (yes/no) signal known exactly (SKE) detection experiment, utilizing 50% probability of signal detection and an ideal observer false alarm rate of 1%, resulted in SNRs of 2.3 (7.23 dB) and 3.6 (11.13) for 50% and 90% true-positive rates, respectively [27].

More recently, the ISO 12232-2006 standard defines two "noise based speeds" for low noise exposures that is based on objective correlation to subjective judgements of the acceptability of various noise levels in exposure series images [28]. The two noise based speeds are defined as $S_{\text{noise}_{40}} = 40, (32.04\,\text{dB})$ providing "excellent" quality images, and $S_{\text{noise}_{10}} = 10, (20\,\text{dB}$ providing "acceptable" quality images. Examining Fig. 2, a commonality point between the ISO 12232-2006 standard and Imatest occurs at the 20-dB point, pertaining to "acceptable" quality in the ISO standard and "high" quality in Imatest. We note that the ISO standard is based on prints at approximately 70 pixels/cm, at a standard viewing distance of 25 cm. For subjective models utilizing sensor pixels per centimeter values P differing from the ISO standard, an SNR scaling ratio of 70/P is recommended.

Previously reported results for objective calibrated DR test data have yielded limited results and the test methodology used is often unclear. The Cinema5D test website [29] notes "We have at one point in 2014 updated our dynamic range evaluation scale to better represent usable dynamic range

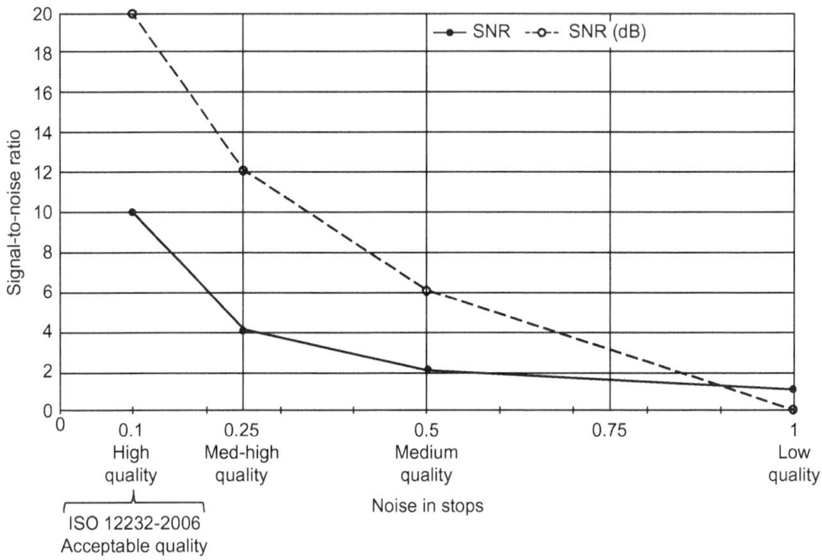

Fig. 2 Arbitrary quality descriptions and SNR in Imatest.

among all tested cameras. This does not affect the relation of usable dynamic range between cameras," however the author does not clearly state what evaluation scale is currently, or was previously, used. In another example, published on the DVInfo website [30], the author describes the minimum input luminance level as "typically, where you consider the noise amplitude to be significant compared to the amplitude of the darkest stop visible above the noise," but does not further define the term "significant." Results published on the DxOMark website [31] primarily include still camera and mobile phone tests, implementing a self-designed DR test chart that is limited to 4 density steps (13.3 stops).

2. METHOD AND MATERIALS

2.1 Design

Overall, the procedure for measuring system dynamic range involves four primary steps. Step one includes the setup of the test environment so that overall lighting conditions can be controlled, with reflected light minimized. The ability to set, maintain, and read room temperature is also of importance, as camera noise may be temperature-dependent. The second step is the setup and alignment of the camera under test

(CUT) and HDR test chart in order to meet the pixels per patch test requirement. With setup activities complete, the third step includes the collection of sample image frames, followed by postprocessing to create compatible input frames for the analysis software. Finally, step four includes processing of the input frames through the analysis software and the review of data.

Three main components in the image capture pipeline that should be considered in measurement of DR include the optics, the sensor, and the camera processing workflow [32]. The optics are important in transforming the scene luminance to the image luminance at the sensor, and can reduce the DR as a result of diffraction, aberration, defocus, and flare [33]. The sensor can have limiting factors to DR as well, including saturation on the intensity high end, and noise limitations on the low end [2]. The camera data processing workflow begins with the color transformation (for single sensor systems), generally utilizing a Bayer color pattern followed by a luminance estimate as a weighted sum of the RGB values. Once the data is recorded, further processing may be required based on proprietary codes, log encoding, or other in-camera data manipulation. Care is taken to postprocess the collected data into a standard format for analysis. While the effects on DR of each of the three main components in the image capture pipeline can be analyzed in detail, a goal of this chapter is to focus on complete system capability, as may be used "out of the box." The measurement and calculation, therefore, includes the combined effects of the optics, sensor, and camera processing as illustrated in the system level model of luminance conversion in Fig. 3.

2.1.1 Test Chart and Processing Software

A DSC Labs Xyla 21 test chart was procured that includes 20 stops of dynamic range via a 21-step, rear-lit, voltage-regulated light source. The Xyla chart, shown in Fig. 4, features a stepped xylophone shape, minimizing flare interference from brighter steps to darker steps. The Xyla chart includes an optional shutter system that allows for the isolation of individual

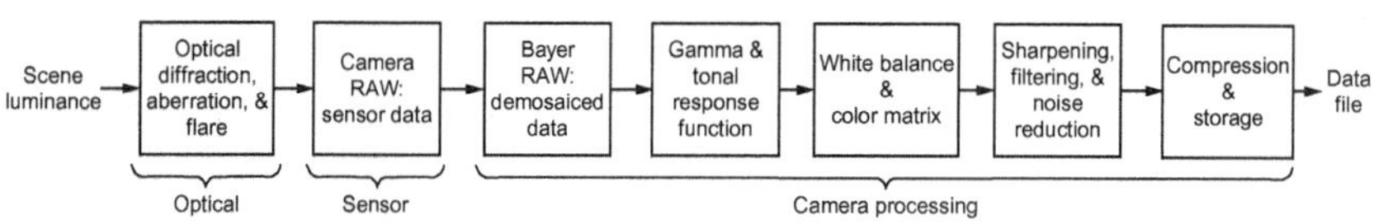

Fig. 3 System level model of luminance conversion.

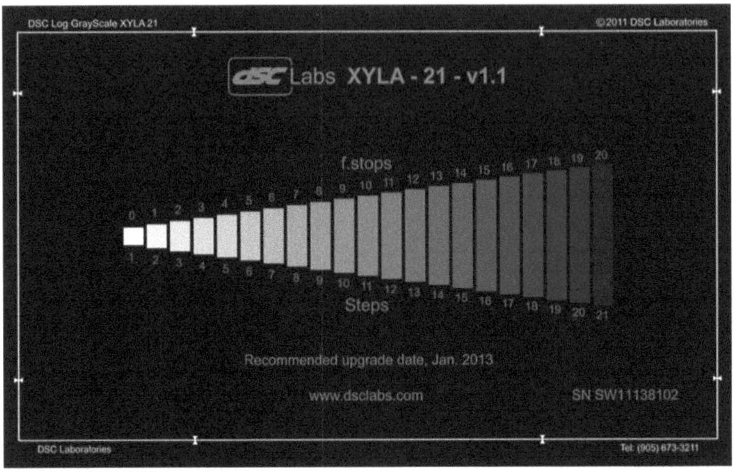

Fig. 4 DSC labs Xyla-21 test chart.

steps in order to reduce the effect of stray light. The use of the shutter system comes at the expense of increased measurement and processing time, as each individual step must be imaged and processed, as opposed to the chart as a whole. The shutter system was not utilized in these experiments due to time constraints; however, the evaluation and characterization of the effect of stray light from the chart itself on the noise measurement is important future work. The Xyla chart is preferred to front-lit reflective charts, as front-lit charts are more difficult to relight evenly over time, as test configurations change or the test environment is altered. Rear-lit, grayscale, stand-alone films require the use of an illuminator, such as a light box, where special care is required to monitor the light source and voltage. The Xyla 21, being an all-enclosed calibrated single unit, simplifies test setup and measurement.

The Xyla-21 chart includes $\log_2$ spaced steps, where each linear density step of 0.3 equates to one stop, a doubling (or halving) of luminance. The calibrated density data for the Xyla-21 chart used is provided in Table 1. The calibration is provided by the manufacturer, and the recommended upgrade (recalibration) date is printed on the chart.

The Stepchart module of the image analysis software Imatest [34] is utilized to process the captured images of the Xyla-21 test chart. The camera distance to the chart is adjusted to maintain approximately 50 pixels per patch horizontal resolution, required by Stepchart [22]. The Stepchart

Table 1 Xyla-21 density calibration data

Step #	**1**	**2**	**3**	**4**	**5**	**6**	**7**	**8**	**9**	**10**	**11**
Density	0	0.3	0.6	0.9	1.2	1.51	1.81	2.11	2.41	2.71	3.01
Step #	12	13	14	15	16	17	18	19	20	21	
Density	3.31	3.61	3.91	4.21	4.52	4.82	5.12	5.42	5.72	6.02	

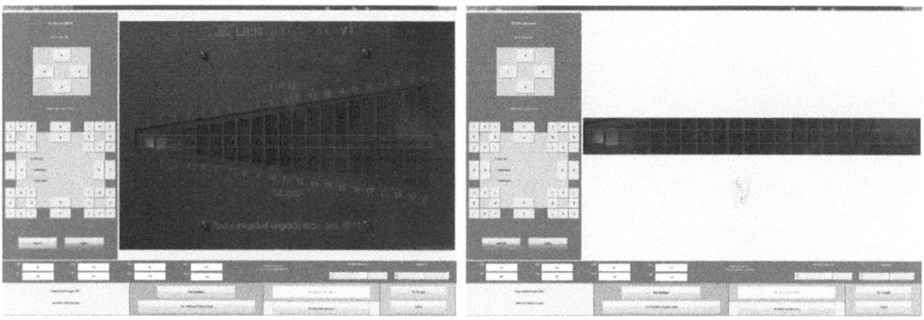

Region of interest selection in alignment window   Zoom and lighten functions enabled

Fig. 5 Imatest Stepchart module region of interest selection screens.

module includes a region of interest (ROI) selection tool that allows for the alignment and selection of ROI for zones as illustrated in Fig. 5. The light gray boxes are the selection region for each individual patch, with software controls allowing adjustment of all patches contained within the red outer box. The right image in Fig. 5 shows the selection tool with zoom and lighten functions turned on, which can be especially helpful when aligning darker zones. A useful alignment hint is to first align the ROIs based on a brightly lit captured frame of the test chart. Imatest will recall this alignment as the default for future images, therefore as long as the camera and chart are not moved, the brightly lit alignment can be used for test frames where the ambient lights have been switched off and darker zones are more difficult to align to. With zone regions selected, Imatest software calculates statistics for each ROI, including the average pixel level and variations via a second-order polynomial fit to the pixel levels inside each ROI. The second-order polynomial fit is subtracted from the pixel levels that will be used for calculating noise (standard deviation), removing the effects of nonuniform illumination appearing as gradual variations.

2.1.2 Selected Cameras

Camera selection was partially based on availability, while best effort was made to include units from various "market" segments. The RED EPIC and ARRI Alexa represent high-end, primarily entertainment industry cameras, both of which claim extended dynamic range. The RED ONE is a very popular entertainment industry camera that continues to find wide use and is available in M and MX models, dependent on the version of the image sensor. The Toshiba and Hitachi cameras represent machine vision cameras that are often used in scientific and engineering applications. The Canon 5D Mark 3 is a latest generation DSLR, and was tested utilizing the manufacturer provided embedded software, as well as open source embedded software known as Magic Lantern [35]. Finally, one of the many new entries into the lower-cost "professional" 4K market cameras, the Black Magic Cinema 4K, was also selected.

2.1.3 Test Procedure

The CUT is mounted on a tripod and placed in a dark test room, kept at an ambient temperature of 23±2°C, with no light leakage along with the Xyla test chart. Camera lights and all reflective surfaces are masked to prevent reflections off the chart surface. A 25-mm Prime PL mount lens is attached to the CUT utilizing PL type mounts. For large format sensors, in order to maintain the 50 pixels per patch horizontal resolution, a longer focal length lens may be required, such as an 85 mm for the RED Dragon 6K. Cameras with nonPL mounts (such as C, Canon EF) utilize available compatible lenses.

A one-foot straight edge is placed across the front plane of the lens. Measurements are taken at each end of the straight edge to the face of the test chart to ensure the two planes are parallel to an eighth of an inch. The lens aperture is set to full open, and focus is set by temporarily placing a Siemens star on the face of the test chart. Once focus has been verified, the camera is set to the manufacturer recommended native ISO, and data collection begins. The ambient room lights are turned off and the door to the test room is closed, so that only the light from the test chart reaches the CUT. The lens remains at the lowest stop value (wide open) to ensure the luminance of at least the first 2–3 chips will saturate the image sensor at the slowest exposure time (generally 1/24th or 1/30th of a second). A few seconds of video are collected, and then the exposure time is reduced by one-half (one stop). Again a few seconds of video are collected, and the process is repeated for a total of at least 5 measurements to complete the sample collection process. Note that only a short video sequence is required, as only a single frame is used for the simultaneous calculation. Imatest does include a sequential (temporal) noise calculation, where additional frames could be utilized in the difference calculation, but the investigation of this will be future work.

Measured frames are processed through Imatest that can be configured to produce output figures relating to exposure, noise, density response, and/or SNR. A graph of normalized exposure pixel level (normalized to 0–255) is one output that is plotted against the step value of the Xyla21 chart. The original image data, whether 8- or 16-bit pixels, is first converted to quad precision (floating point) and normalized to 1 before any processing is performed. The normalization to 255 is only for the purposes of the output graphical display (historically related to 8-bit file depth and the familiar 255 maximum pixel level). Therefore it is important to note that the full precision of the input file data is utilized in the calculations, independent of the normalized output scale. Sample normalized exposure data for the ARRI Alexa is shown in Fig. 6, noting that seven exposures were collected during this test. The collection of multiple measurements is to ensure an exposure was captured meeting the criteria of the maximum unclipped input luminance level L_{sat} defined in Imatest as 98% of its maximum value.

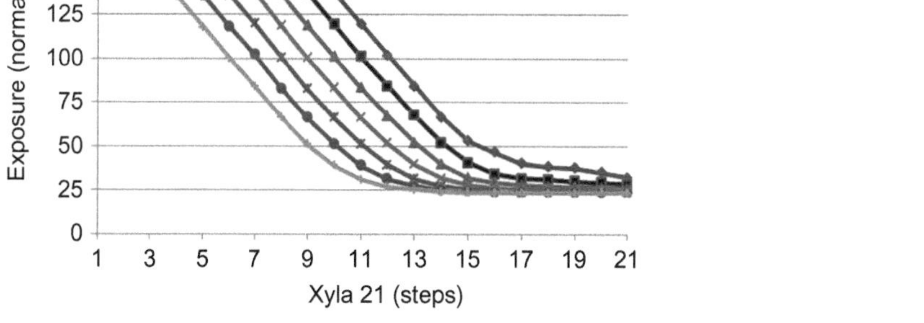

Fig. 6 Normalized exposure data for ARRI Alexa.

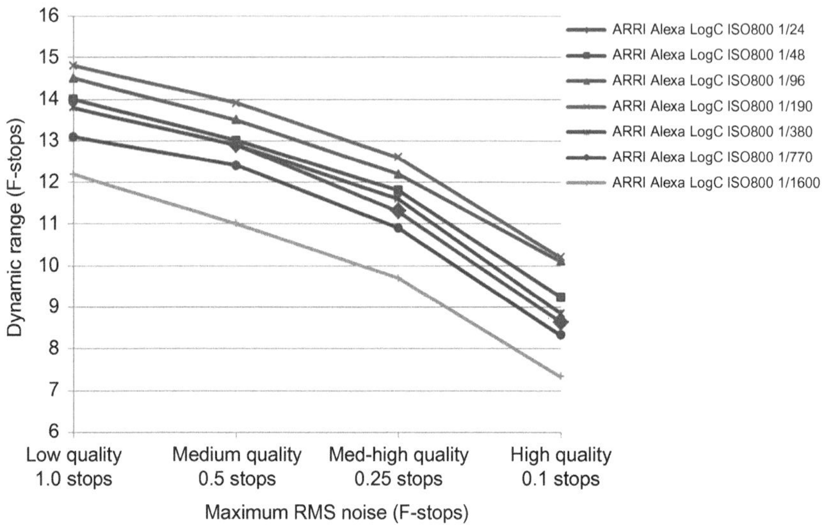

Fig. 7 Dynamic range data for ARRI Alexa.

DR is calculated by Imatest, using the minimum luminance step, meeting the maximum RMS noise for each exposure time. This is shown in Fig. 7 for the ARRI Alexa. We can observe the exposure setting that resulted in the peak dynamic range was the one where the brightest chip was just saturated. In the case of the ARRI Alexa data, this was the exposure time setting of 1/190 of a second, resulting in 13.9 stops for a maximum 0.5 stop RMS noise utilized in defining the minimum luminance step.

The RED EPIC and DRAGON HDR modes are treated similarly, however, an estimate of the total dynamic range is made based on a combination of both the A and X frames. HDR exposure data for the RED EPIC is shown in Fig. 8, including a "shifted" X frame normalized to the sixth A frame step. Note that for both the RED EPIC and DRAGON, the longest exposure data was utilized to ensure saturation of the A frame, so that more of the X frame is utilized. There may be a limitation in the current setup, where the peak absolute luminance of the Xyla chart is not bright enough to fully extend into the X frame range.

2.1.4 Data Workflow

A significant challenge in testing a large sample of digital imagers is developing data workflows. For as many vendors utilized, there are just as many proprietary data formats, encoding algorithms, export formats, software packages, and postprocessing guidelines. How captured digital

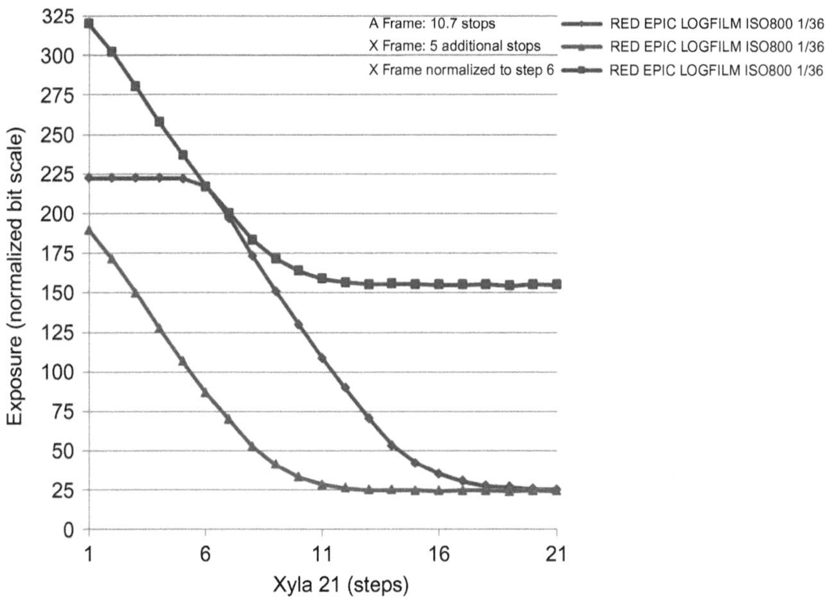

Fig. 8 Normalized exposure data for RED EPIC A and X frame.

data is provided to the user is also manufacturer dependent. For example, ARRI utilizes a dual gain architecture (DGA) with two paths of different amplification for each pixel [36]. ARRI processes the dual path data internal to the camera, providing a single output file, available either as RAW (with external recorder), LogC, or Rec709. ARRI file formats typically use ARRI LUTs for gamma and color space, but are wrapped in standard ProRes422 or ProRes444 wrappers. Generation and application of ARRI specific LUTs for the different outputs is straightforward, resulting in a user-friendly workflow [37].

In some cases, file wrappers, encoding schemes, gamma, and color space functions are vendor-specific, such as with RED Redcode files, but have wide acceptance and are importable and manageable, either through vendor provided software or popular third-party applications [38]. RED provides its HDR digital data as two separate file sequences comprising of A and X frames. Access to both exposure frames has the benefit of increased flexibility in postprocessing, at the cost of increased complexity in understanding the tone-mapping process. The RED software CineXPro includes two tone-mapping functions, Simple Blend and Magic Motion, to assist the user in the tone-mapping process [39].

Table 2 System level model of luminance conversion

Imager	Capture format	Processing steps
ARRI Alexa	12 Bit LogC, ProRes444, 1920×1080	LogC—Export 16 Bit TIFF from DaVR Rec709—Apply 3D LUT, export 16 Bit TIFF from DaVR
RED One M/MX	12 Bit Linear, Redcode36, 4096×2034	Log/LogFilm Gamma—Apply using CineXPro, export 16 Bit TIFF Gamma3/Color3—Apply using CineXPro, export 16 Bit TIFF
RED EPIC/DRAGON	12 Bit Linear, Redcode 8:1, EPIC 4K 16:9, DRAGON 6K 16:9	LogFilm Gamma—Apply using CineXPro, export 16 Bit TIFF Apply Simple Blend, Magic Motion tone-mapping as required.
Toshiba HD Hitachi DK	Uncompressed YUV, Blackmagic Hyperdeck Pro Recorder, .MOV wrapper	Rec709 gamma applied by camera. Export 16 Bit TIFF from DaVR
BlackMagic	Film Mode, Apple ProRes 422HQ, .MOV wrapper, 3840x2160	Film gamma applied by camera. Export 16 Bit TIFF from DaVR
Canon 5DM3 Magic Lantern RAW Still	ML RAW, Full Frame	Convert ML RAW using RAW2DNG. Export 16 bit TIFF DaVR
Canon 5DM3 Magic Lantern HDR Movie	ML H.264, 1920×1080	Convert H.264 HDR file using AVISynth to HDR JPEG.

Manufacturer specific postprocessing workflow and applications used in the creation of sample frames for analysis are summarized in Table 2.

3. RESULTS

Results of imagers tested are shown in Table 3, including the native ISO values used and the Vendor specification when available. The dynamic

Table 3 Dynamic range measurement of digital imagers

Imager—mode, gamma	Stops @ 0.1 RMS noise	Stops @ 0.25 RMS noise	Stops @ 0.5 RMS noise	Stops @ 1.0 RMS noise	Native ISO	Vendor spec.
Quality term	**High**	**Medium-high**	**Medium**	**Low**		
RED DRAGON— HDR x6, Log Film, Total	7.6 + ~ 3	10.2+~ 3	11.8 + ~ 3	13.1 + ~ 3	250	16.5+
RED EPIC—HDR x6, Log Film, Total	7.6 + ~ 5	9.2 + ~ 5	10.7 + ~ 5	12.0 + ~ 5	800	< 18
RED EPIC—HDR x6, Log Film, Simple Blend	11.6	13.8	15.0	16.1	800	N/A
ARRI Alexa—Log	10.2	12.6	13.9	14.8	800	14.0
RED EPIC—HDR x6, Log Film, Magic Motion	10.0	11.7	13.1	14.2	800	N/A
RED DRAGON—HDR Off, Log Film	8.4	11.1	12.4	13.6	250	N/A
ARRI Alexa—Rec709	8.2	10.7	11.8	12.7	800	N/A
RED EPIC—HDR Off, Log Film	8.6	10.6	11.5	12.8	800	13.5
RED ONE MX—Log Film	8.8	10.4	11.5	12.7	800	13.0+
RED ONE MX—Gamma 3	8.5	10.4	11.4	12.2	800	13.0+
Canon 5DM3—Magic Lantern H.264 HDR DUAL ISO 400/1600	8.5	10.5	11.4	11.9	100	14.0
Toshiba IK-HR1S HD—Gamma 0	8.5	9.5	11.1	12.0	0 dB	N/A
Canon 5DM3—RAW Still	7.6	8.8	10.6	11.6	100	N/A
Hitachi DK-H100—Gamma On	3.9	8.3	10.6	11.2	-3^{a}	10.0+
RED ONE M—Log	7.7	8.9	10.4	11.7	320	11.3
RED One M—Gamma 3	7.8	9.3	9.9	10.6	320	N/A
Hitachi DK-Z50—Gamma On	5.3	7.2	9.3	10.1	-3^{a}	9.6
Black Magic Cinema 4K—Film	6.0	7.8	9.0	10.0	200	12.0

[a] Hitachi cameras do not specify ISO values for digital gain, instead an arbitrary gain of −3, 0, or +3 is available.

range measurements are stated in terms of the minimum luminance step meeting the maximum RMS noise, with high, medium-high, medium, and low quality corresponding to 0.1, 0.25, 0.5, and 1.0 stops of RMS noise, respectively. The RED EPIC and DRAGON total measurements are approximate, as it includes the combination of the two measured A and X frames. Data for the RED tone-mapping functions Simple Blend and Magic Motion have been included, as well as measurements for the RED EPIC and DRAGON without utilizing the HDR function.

In Table 3 results, we attempted to select a "most appropriate" single quality term, and accompanying maximum RMS noise defining the minimum luminance step, that correlates with published data. The table has been arranged from highest to lowest DR, using the Imatest designated "medium" quality. Sorting of the data based on the medium quality term is made after comparison to available Vendor specifications provide where available. The high and medium-high quality results appear to be too stringent to be utilized by most Vendors, especially when marketing for typical applications. Medium quality has relatively close agreement when considering high-end cameras, such as the ARRI Alexa (measured 13.9 stops, specified 14 stops) and industrial cameras, such as Hitachi DK-H100 (measured 10.6 stops, specified 10.0+ stops) and DK-Z50 (measured 9.3 stops, specified 9.6 stops). The case could be made for utilizing the low quality term, as measured data does have close agreement to Vendor specifications in the case of the RED ONE M Log (measured 11.7 stops, specified 11.3 stops), RED ONE MX Log (measured 12.7 stops, specified 13.0+ stops), and Black Magic 4K camera (measured 10.0 stops, specified 12.0 stops). This option seems less than ideal however, as higher noise levels in dark regions are generally undesirable and the low quality term includes a full stop of RMS noise.

4. DISCUSSION

An evaluation of a NROQM to accurately measure and compare the increased dynamic range capability of modern image capture systems is presented. The method utilizes a 21-step, back lit test chart providing luminance patches over a 20-stop range. Data collection, processing, and analysis steps are described as part of an image capture and postprocessing workflow. The workflow requires the use of Vendor software, commercially available postprocessing software, or in some cases, both. Sample data is presented for a number of current generation imaging systems, some of which are extending DR capability beyond traditional digital imaging capabilities.

The image systems tested illustrate the different technologies and implementations available in the market today. In some cases, Vendors utilize a single sensor with single output, relying on the intrinsic capability of the sensor for the maximum DR. In other cases, such as the ARRI Alexa, dual gain sensors are implemented with the Vendor combining, or tone-mapping the dual data sets into a single output, transparent to the end user. Furthermore, with devices such as the RED EPIC and DRAGON, multiple exposures are taken and stored as separate data fields, leaving the tone-mapping as an additional step in the postprocessing workflow. An observation that can be made from testing these various images systems is that when considering DR, it is equally important to consider the postprocessing workflow and how it will be conducted. Manufacturers also offer various gamma functions, which when utilized, result in changes in the final DR.

In real world scenes, quality depends on *scene* content, such as edges and gradients, as well as target contrast of a signal with respect to the background. Ultimately, we can state that arbitrary quality descriptors are best described as scene dependent. What is deemed "excellent" for one application, or scene, may be "acceptable" for another. Producers of imagery for entertainment may not only allow, but welcome more low light noise to achieve a film like appearance or "look." Other users, such as in the medical, engineering, or scientific fields, may alternatively require increased signal to noise performance. In other cases, the human eye has been shown to detect well-defined targets when the SNR is less than 0.1 [10]. Manufacturers often add to the confusion by publishing DR or minimum signal level without reference to test conditions or SNR, or without stating if the data is presented for the ideal sensor case, making comparisons difficult. We can state the obvious regarding the definition and application of arbitrary descriptive terms, that they are in fact arbitrary, but can have value and can be compared when they are characterized with respect to a meaningful metric, such as the SNR of the minimum luminance level used to determine DR. The medium quality level pertaining to a 0.5 RMS noise was chosen as a compromise between results that correlate with Vendor supplied data, and a noise level that is undesirable. Medium quality was selected for the general case, however it is open for interpretation for specific use cases. Future work will be conducted in correlating RMS noise based quality evaluation with subjective evaluations.

The results indicate that not all Vendor data correlates with a specific maximum RMS noise defining the minimum luminance step. This may be due to differences in test methodologies, to acceptance of different noise levels, or both. Also of note is the higher end ARRI and RED imaging systems resulted in the greatest DR, however, the RED EPIC and DRAGON results for HDR mode are estimates based on the separate exposure frames. The final realized DR of the RED HDR systems will be a factor of the tone-mapping method employed.

5. CONCLUSION

As imaging systems continue to develop and evolve, including recent growth in HDR, calibrated measurements of DR that are both transparent and comparable are of increased importance. Manufacturer specifications are often without reference or descriptive narrative as to the computational method. The use of arbitrary quality descriptors may indeed be arbitrary, as what is deemed "excellent" for one application, or scene, may be "acceptable" for another. Value can be added to dynamic range measurement and quality descriptors, and they can be better compared, when characterized with respect to a meaningful metric, such as the maximum RMS noise defining the minimum luminance step used to determine DR. A SNR of 20 dB (0.1 stop of RMS noise) was identified as a reference point having commonality to both ISO Standard 12232-2006 and the test software Imatest. Review of measured results for several modern imaging systems indicate that the industry trend in DR reporting correlates closer with the "medium" quality descriptor in Imatest corresponding to a SNR of 6 dB (0.5 stop of RMS noise when defining the minimum luminance patch), than with the 20 dB metric termed "acceptable" quality in the ISO Standard 12232-2006 and "high" quality in Imatest. The imagers tested have been treated as "systems," with the combined effect of the optics, the sensor hardware, and the camera processing. Future work will include more specific analysis of the effects on DR based on the contribution of each of the individual sub-systems.

ACKNOWLEDGMENTS

Customer support was provided by RED Inc., ARRI Inc., Imatest LLC, and DSC Laboratories. Debattista and Chalmers are partially funded as Royal Society Industrial Fellows. This project is also partially supported by EU COST Action IC1005.

REFERENCES

[1] F. Banterle, A. Artusi, K. Debattista, A. Chalmers, Advanced High Dynamic Range Imaging: Theory and Practice, AK Peters, Natick, MA, 2011.
[2] E. Reinhard, W. Heidrich, P. Debevec, S. Pattanaik, G. Ward, K. Myszkowski, High Dynamic Range Imaging: Acquisition, Display, and Image-Based Lighting, Elsevier Science, Amsterdam, 2010.
[3] T.N. Pappas, R.J. Safranek, J. Chen, Perceptual criteria for image quality evaluation, in: Handbook of Image and Video Processing, Bell Laboratories – Lucent Technologies, Murray Hill, NJ, 2000, pp. 669–684.
[4] ITU, Methodology for the subjective assessment of the quality of television pictures, 2012, ITU-R.
[5] M. Narwaria, M. Perreira da Silva, P. Le Callet, HDR-VQM: an objective quality measure for high dynamic range video, Signal Process. Image Commun. 35 (2015) 46–60.
[6] M. Narwaria, R.K. Mantiuk, M.P. da silva, P. Le Callet, HDR-VDP-2.2: a calibrated method for objective quality prediction of high-dynamic range and standard images, J. Electron. Imaging 24 (2015) 010501.
[7] G. Valenzise, F. de Simone, P. Lauga, F. Dufaux, Performance evaluation of objective quality metrics for HDR image compression, in: Applications of Digital Image Processing XXXVII, 92170C, SPIE, 2014, 10 p.
[8] M. Carnec, P. le Callet, D. Barba, Objective quality assessment of color images based on a generic perceptual reduced reference, Signal Process. Image Commun. 23 (2008) 239–256.
[9] H.R. Wu, K.R. Rao, A no-reference quality metric for degraded and enhanced video, in: Digital Video Image Quality and Perceptual Coding, CRC Press, Boca Raton, FL, 2005.
[10] G.C. Holst, T.S. LOMHEIM, CMOS/CCD Sensors and Camera Systems, JCD Publishing, Bellingham, WA, 2007.
[11] ISO, Photography. Electronic still-picture imaging. Noise measurements, 2013, BSI Standards Limited.
[12] C.A. Poynton, Digital Video and HDTV: Algorithms and Interfaces, Morgan Kaufmann Publishers, San Francisco, CA, 2003.
[13] C. Ma, Y. Liu, J. Li, Q. Zhou, Y. Chang, X. Wang, A 4MP high-dynamic-range, low-noise CMOS image sensor, in: International Society for Optics and Photonics, vol. 940305, IS&T/SPIE Electronic Imaging, 2015.
[14] D.L. Cade, Fuji and Panasonic's New Organic Sensor Boasts Insane 29.2 Stop Dynamic Range, 2013, Available: http://petapixel.com/2013/06/11/fuji-and-panasonics-new-organic-sensor-boasts-insane-14-6-stops-dynamic-range/ (accessed 08.10.14).
[15] T. Kunkel, E. Reinhard, A reassessment of the simultaneous dynamic range of the human visual system, Proceedings of the 7th Symposium on Applied Perception in Graphics and Visualization (APGV'10), ACM, Los Angeles, CA, 2010.
[16] J.A. Ferwerda, Elements of early vision for computer graphics, IEEE Comput. Graph. 21 (2001) 22–33.
[17] B.A. Wandell, Foundations of Vision, Sinauer Associates, Sunderland, MA, 1995.
[18] Y. Reibel, M. Jung, M. Bouhifd, B. Cunin, C. Draman, CCD or CMOS camera noise characterisation, Eur. Phys. J. Appl. Phys. 21 (2003) 75–80.
[19] R.D. Fiete, Modeling the Imaging Chain of Digital Cameras, SPIE Press, Washington, 2010.
[20] G.E. Healey, R. Kondepudy, Radiometric CCD camera calibration and noise estimation, IEEE Trans. Pattern Anal. Mach. Intell. 16 (1994) 267–276.
[21] European Machine Vision Association, Standard for Characterization of Image Sensors and Cameras, 2010, European Machine Vision Association, *Release*.

[22] N. Koren, Digital Image Quality Testing Software, 2014, Imatest LLC., Available: http://www.imatest.com/.
[23] R. Mantiuk, G. Krawczyk, H.P. Seidel, R. Mantiuk, High dynamic range imaging pipeline: perception-motivated representation of visual content, in: Proceedings of SPIE-IS&T Electronic Imaging, SPIE, 2007/01/01/2007.12.
[24] A. Rose, The sensitivity performance of the human eye on an absolute scale, J. Opt. Soc. Am. 38 (1948) 196–208.
[25] J.A. Nevin, Signal detention theory and operant behavior: a review of David M. Green and John A. Swets' signal detection theory and psychophysics, J. Exp. Anal. Behav. 12 (1969) 475–480.
[26] D. Green, J. Swets, Signal Detection Theory and Psychophysics, Peninsula Publishing, 1966.
[27] A.E. Burgess, The Rose model, revisited, J. Opt. Soc. Am. A 16 (1999) 633–646.
[28] ISO, Photography. Digital still cameras. Determination of exposure index, ISO speed ratings, standard output sensitivity, and recommended exposure index, 2006, BSI Standards Limited.
[29] S. Wober, Dynamic Range—Sony A7S vs. the others, 2014, Cinema5D e.U., Available: https://www.cinema5d.com/dynamic-range-sony-a7s-vs-arri-amira-canon-c300-5d-mark-iii-1dc-panasonic-gh4/ (accessed 08.09.15).
[30] A. Wilt, Panasonic GH4: DR and Gamma, Time lapse, and More, 2014, DV Info Net., Available: http://www.dvinfo.net/article/acquisition/micro-43/panasonic-gh4-dr-and-gamma-timelapse-and-more.html (accessed 08.09.15).
[31] DxOmark, Testing protocols for noise and dynamic range, 2015, DxO Labs, Available: http://www.dxomark.com/About/In-depth-measurements/DxOMark-testing-protocols/Noise-dynamic-range (accessed 08.09.15).
[32] F. Xiao, J.M. Dicarlo, P.B. Catrysse, B.A. Wandell, High dynamic range imaging of natural scenes, in: Color and Imaging Conference, 2002, pp. 337–342.
[33] J.J. McCann, A. RIZZI, Camera and visual veiling glare in HDR images, J. Soc. Inf. Disp. 15 (2007) 721–730.
[34] Imatest, Step Chart Module Documentation 2014, Imatest LLC., Available: http://www.imatest.com/support/modules/stepchart/ (accessed 03.02.15).
[35] MagicLantern, Magic Lantern Overview 2014, MagicLantern, Available: http://www.magiclantern.fm (accessed 10.08.15).
[36] ARRI, Alexa's Sensor, 2013, ARRI Inc., Available: http://www.arri.com/camera/alexa/technology/arri_imaging_technology/alexas_sensor (accessed 26.08.14).
[37] ARRI, Alexa LUT Generator: LookUp Tables 2013, ARRI Inc., Available: http://www.arri.com/camera/alexa/tools/lut_generator/ (accessed 10.08.14).
[38] RED, Understanding RED LogFilm and RedGamma, 2013, Red Inc., Available: http://www.red.com/learn/red-101/redlogfilm-redgamma (accessed 25.08.14).
[39] RED, REDCINE-X Post Production Software Overview, 2014, Red Inc., Available: http://www.red.com/products/redcine-x-pro/ (accessed 15.11.14).

PART 2

Manipulate

CHAPTER 5

HDR Video Metrics

M. Čadík*, T. Ozan Aydın†
* Brno University of Technology, Brno, Czech Republic
†Disney Research, Zürich, Switzerland

Contents

1. INTRODUCTION

In this chapter we present an introduction to HDR image and video quality assessment fields. We discuss full-, no-, and reduced-reference metrics, including perceptually motivated methods. We describe two existing full-reference HDR video quality metrics in detail. Furthermore, we introduce the emerging field of data-driven metrics. Finally, we conclude with the outlook of future development and research.

2. IMAGE AND VIDEO QUALITY ASSESSMENT

The goal of image and video quality assessment (IQA, VQA) is to computationally predict human perception of image and video quality. Practical evidence shows [1, 2] that numerical distortion metrics, like root mean squared error (RMSE), are often not adequate for the comparison of

High Dynamic Range Video
http://dx.doi.org/10.1016/B978-0-12-809477-8.00005-4

images, because they poorly predict the differences between the images as perceived by a human observer. To solve this problem properly, various *image and video quality metrics* (IQM, VQM) have been proposed [2]. IQM traditionally comprise a computational human visual system (HVS) model to correctly predict image difference as a human would perceive it, rather it is a bottom-up [3], or a top-down approach [4]. Please refer to vision science textbooks [5] for an in-depth treatment of human visual perception, and on HVS measurements related to masking, adaptation, contrast sensitivity, etc.

Image and video quality assessment is practical in various applications. The main applications of IQA lie in the areas of image quality monitoring (e.g., in lossy image compression), benchmarking of imaging applications, and optimizing algorithms by tuning their parameter settings. Furthermore, IQM have also been successfully applied to image database retrievals, or evaluation of the perceptual impact of different computer graphics and vision algorithms.

In the following text, we will survey existing standard dynamic range (SDR) quality assessment approaches, while the only two existing metrics designed specifically for HDR video processing will be described in Sections 3 and 4 in detail.

2.1 Full-Reference Metrics

Full-reference IQM and VQM are based on measuring the errors (signal differences) between a distorted image and the reference image. The aim is to quantify the errors in a way that simulates human visual error sensitivity. Video quality assessment is often inspired by the ideas from the more developed image quality assessment field. A large variety of SDR IQM have been proposed in the literature [1, 2]. Traditionally, IQM focus on near-threshold detection [6], suprathreshold discrimination [7], or functional differences [8].

Video metrics usually extend IQM with temporal models of visual perception, resulting from the fact that frame-by-frame application of IQM is not sufficient. Van den Branden Lambrecht's moving picture quality metric (MPQM) [9] utilizes a spatial decomposition in frequency domain using a filter bank of oriented Gabor filters, each with one octave bandwidth. Additionally two temporal channels, one low-pass (sustained) and another band-pass (transient) are computed to model visual masking. The output of their metric is a numerical quality index between 1 and 5, similar to the mean opinion score obtained through subjective studies. In a more efficient version of MPQM, the Gabor filter bank is replaced by the steerable pyramid [10]. In later work targeted specifically to assess the quality of MPEG-2 compressed videos [11], they address the space-time nonseparability of contrast sensitivity through the use of a spatiotemporal model. Another metric based on steerable pyramid decomposition aimed toward low bit-rate videos with severe artifacts is proposed by Masry and Hemani [12], where they use finite impulse response filters for temporal decomposition.

Watson et al. [13] proposed an efficient digital video quality metric (DVQ) based on the discrete cosine transform. The DVQ models early HVS processing, including temporal filtering and simple dynamics of light adaptation and contrast masking. Later they propose a simple standard spatial observer (SSO) based method [14], which, on the Video Quality Experts Group data set, is shown to make as accurate predictions as more complex metrics. Winkler [15, 16] proposed a perceptual distortion metric (PDM) where he introduced a custom multiscale isotropic local contrast measure that is later normalized by a contrast gain function that accounts for spatiotemporal contrast sensitivity and visual masking.

A video metric V-SSIM proposed by Seshadrinathan and Bovik [17] is an extension to the IQM called complex wavelet structural similarity index (CW-SSIM) [18, 19], to account for motion in video sequences. The technique incorporates motion modeling using *optical flow* and relies on a decomposition through 3D Gabor filter banks in frequency domain. V-SSIM is therefore able to account for motion artifacts due to quantization of motion vectors and motion compensation mismatches. The same authors published the MOVIE index in a follow-up work [20], which outputs two separate video quality streams for every 16th frame of the assessed video: *spatial* (closely related to the structure term of SSIM) and *temporal* (assessment of the motion quality based on optical flow fields).

2.2 No-Reference and Reduced-Reference Metrics

The main issue in developing *no-reference (NR) metrics* is the absence of a nondistorted reference image or some features representing it. Common approaches to compensate for this are (1) modeling distortion-specific characteristics, (2) using natural scene statistics, and (3) employing learning based classification methods.

Distortion-specific NR methods capitalize on the knowledge of artifact type and its unique characteristics [21] (Chapter 3). Examples include metrics for detecting blockiness due to lossy JPEG and MPEG compression, and ringing at strong contrast edges [22], blurriness due to high frequency coefficients suppression [23, 24], and banding (false contouring) at low gradient regions due to the excessive quantization [25]. There are some attempts at building more general NR quality metrics, which evaluate a combined contribution of individually estimated image features such as sharpness, contrast, noise, clipping, ringing, and blocking artifacts [21] (Chapter 10). The contribution of all features, including their simple interactions is summed up with weights derived through fitting to subjective data.

Natural scene statistics [26] derived from artifact-free images can be helpful in detecting artifacts. Sheikh et al. [27] show that noise, blurriness, and quantization can be identified as deviations from these statistics.

Image features extracted from distorted and nondistorted images are used for training machine learning techniques, such as support vector machines (SVM) or neural networks. Moorthy and Bovik [28] use generalized Gaussian distribution (GGD) to parameterize wavelet subband coefficients and create 18-D feature vector (3 scales × 3 orientations × 2 GGD parameters), which is used to train an SVM classifier based on perceptually calibrated distortion examples from the LIVE IQA database. The classifier discriminates between five types of mostly compression-related distortions and estimates their magnitude. Saad et al. [29] train a statistical model to detect distortions in DCT-based contrast and structure features.

Reduced-reference metrics may be beneficial in video compression or transmission monitoring [30–33], where the bandwidth is limited. The challenge is to select a representative set of features, which are extracted from an undistorted signal and transmitted along with the possibly distorted image or video. In their pioneering work, Webster et al. [34] used localized spatial and temporal activity channels for this purpose. Later on, Redi et al. [35] identified the *color correlograms* as suitable feature descriptors for analysis of alterations in the color distribution as a result of distortions.

3. DRI-VQM

The recent proliferation of high dynamic range (HDR) imaging forces video quality assessment metrics to be accurate in extended luminance ranges. This requirement limits the use of legacy VQM designed for detecting compression artifacts in SDR videos. Moreover, applications such as tone mapping and compression of HDR video sequences require detecting structural distortions where the reference video is HDR and the test video is SDR. DRI-VQM is a VQM that is designed specifically for these recently emerged practical problems.

DRI-VQM utilizes an HDR capable human visual system (HVS) model that accounts for both major spatial and temporal aspects of the visual system, and employs a pair of dynamic range independent distortion measures *contrast loss* and *amplification* introduced in its counterpart for images, DRI-IQM [36]. DRI-VQM also computes the *visible differences* between reference and test videos, similar to conventional VQM. In most visual computing and video quality assessment applications, the main concern is often the existence of visible artifacts, rather than the magnitude of visibility. Methods that produce clearly visible artifacts are often not useful in practice. Consequently, DRI-VQM's HVS model trades off suprathreshold precision for accuracy near the detection threshold.

The computational steps of DRI-VQM are summarized in Fig. 1. The input is a pair of videos V_{ref} and V_{tst} with arbitrary dynamic ranges, both of which should contain calibrated luminance values. The luma values of SDR videos should be inverse gamma corrected and converted to display luminance (In all examples, we assumed a hypothetical display with the luminance range 0.1–100 cd/m^2 and gamma 2.2.). The HVS model is then applied separately to both videos to obtain the normalized multi-channel local contrast at each visual channel, where the first step is to model the nonlinear response of the photoreceptors to luminance, namely **luminance adaptation**. In DRI-VQM we apply the nonlinearity which maps the video luminance to linear just noticeable differences (JND) values, such that the addition or subtraction of the unit value results in a just perceivable change of relative contrast.

Contrast sensitivity is a function of spatial frequency ρ and temporal frequency ω of a contrast patch, as well as the current adaptation luminance of the observer L_a. The spatiotemporal CSF^T plotted in Fig. 2C shows the human contrast sensitivity for variations of ρ and ω at a fixed adaptation

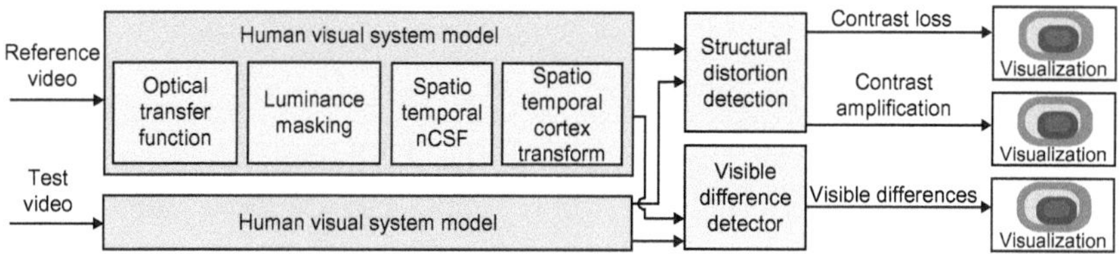

Fig. 1 Data flow diagram of DRI-VQM (see text for details).

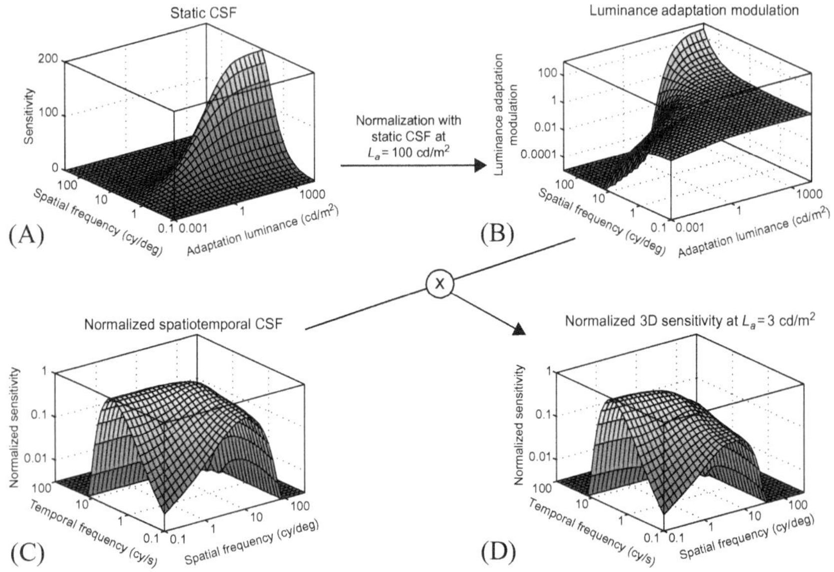

Fig. 2 Computation of the CSF^{3D}. The static $CSF^S(\rho, L_a)$ (A) is divided to $CSF^S(\rho, L_a = 100\,cd/m^2)$ to obtain scaling coefficients (B) that account for luminance adaptation in CSF^{3D}. The specific adaptation level is chosen to reflect the conditions where the spatiotemporal CSF^T was measured (C). The scaling coefficients are computed for the current L_a ($3\,cd/m^2$ in this case), and multiplied with the normalized CSF^T to obtain the CSF^{3D} that accounts for spatial and temporal frequencies, as well luminance adaptation (D).

luminance. At a retinal velocity ν of 0.15 degree/s, the CSF^T is close to the static CSF^S [6] (Fig. 2A) at the same adaptation level (the relation between spatiotemporal frequency and retinal velocity is $\omega = \nu\rho$, assuming the retina is stable). This particular retinal velocity corresponds to the lower limit of natural drift movements of the eye, which are present even if the eye is intentionally fixating in a single position [37]. In the absence of eye tracking data, DRI-VQM assumes that the observer's gaze is fixed, but also the drift movement is present. Accordingly, a minimum retinal velocity is set as follows:

$$CSF^T(\rho, \omega) = CSF^T(\rho,\ \max(\nu, 0.15) \cdot \rho). \qquad (1)$$

On the other hand, the shape of the CSF depends strongly on adaptation luminance, especially for scotopic and mesopic visions, and remains approximately constant over $1000\,cd/m^2$. Consequently, using a spatiotemporal CSF at a fixed adaptation luminance results in erroneous predictions of sensitivity at the lower luminance levels that can be encoded in HDR images. Thus, we derive a "3D" CSF (Fig. 2D) by first computing a *Luminance Modulation Factor* (Fig. 2B) as the ratio of CSF^S at the observer's current adaptation luminance (L_a) with the CSF^S at $L_a = 100\ \mathrm{cd/m}^2$, which is the adaptation level at which the CSF^T is calibrated to the spatiotemporal sensitivity of the HVS. This factor is then multiplied with the normalized spatiotemporal CSF ($n\mathrm{CSF}^{\mathrm{T}}$), and finally the resulting $\mathrm{CSF}^{3\mathrm{D}}$ accounts for ρ, ω, and L_a:

$$\mathrm{CSF}^{3\mathrm{D}}(\rho, \omega, L_a) = \frac{\mathrm{CSF}^S(\rho, L_a)}{\mathrm{CSF}^S(\rho, 100)} n\mathrm{CSF}^{\mathrm{T}}(\rho, \omega). \tag{2}$$

Ideally the $\mathrm{CSF}^{3\mathrm{D}}$ should be derived from psychophysical measurements in all three dimensions, because current findings suggest that the actual contrast sensitivity of the HVS is linearly separable in neither of its dimensions. In the absence of such measurements, estimating luminance adaptation using a scaling factor is better than the alternatives that involve an approximation by linear separation of spatial and temporal frequencies. The effect of luminance adaptation to spatiotemporal contrast sensitivity is approximately linear, except for very low temporal frequencies [38, p. 233].

The perceptually scaled luminance contrast is then decomposed into *visual channels*, each sensitive to different temporal and spatial frequencies and orientations. For this purpose DRI-VQM extends the **Cortex Transform** [39] that comprises six spatial frequency channels, each further divided into six orientations (except the base band), by adding a sustained (low temporal frequency) and a transient (high temporal frequency) channel in the temporal dimension (total 62 channels). The time (t given in seconds) dependent impulse responses of the sustained and transient channels, plotted in Fig. 3A, are given as Eq. (3) and its second derivative, respectively [16]:

$$f(t) = \mathrm{e}^{-\frac{\ln(t/0.160)}{0.2}}. \tag{3}$$

The corresponding frequency domain filters are computed by applying the Fourier transform to both impulse responses and are shown in Fig. 3B.

Combining all models discussed so far, the computation of visual channels from the calibrated input video V is performed as follows:

$$C^{k,l,m} = \mathcal{F}^{-1}\left\{V_{csf}\ \mathrm{cortex}^{k,l} \times \mathrm{temporal}^m\right\},$$
$$V_{csf} = \mathcal{F}\{jnd(V)\}\ \mathrm{CSF}^{3\mathrm{D}},$$

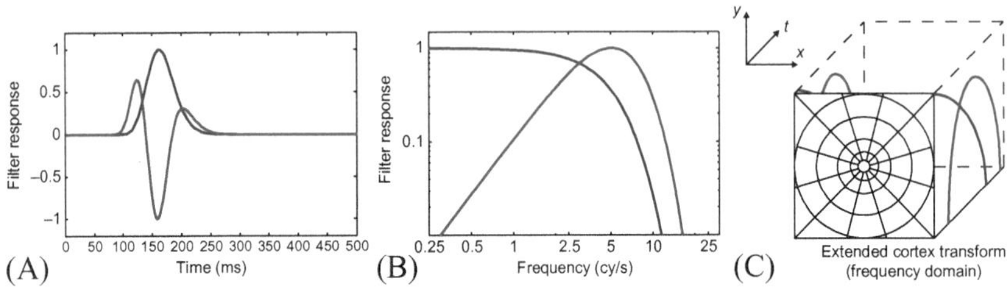

Fig. 3 Impulse (A) and frequency (B) responses of the transient (red) and sustained (blue) temporal channels. The frequency responses comprise the extended 3D Cortex Transform's channels in temporal dimension (C).

where the 3D Cortex filter for channel $C^{k,l,m}$ is computed from the corresponding 2D cortex filter cortex^{k,l} at spatial frequency level k and orientation l, and the sustained and transient channel filters temporal^{m}. The function *jnd* denotes the light adaptation nonlinearity, and $\mathscr{F}$ is the Fourier Transform.

The detection probability of the normalized contrast response C at each visual channel is computed using the following **psychometric function** separately for the reference and test images:

$$P(C) = 1 - \exp(-|C|^3). \tag{4}$$

We compute the probability of detecting a **visible difference** between videos ($P(C_{\text{tst}} - C_{\text{ref}})$), as well as two dynamic range independent distortion measures from individual detection probabilities of the contrast in visual channels [36]. The per-channel dynamic range independent distortion measures are defined as follows:

- **Contrast loss** $\left(P_{\searrow}^{k,l,m} = P(C_{\text{ref}}^{k,l,m})(1 - P(C_{\text{tst}}^{k,l,m})\right)$
- **Contrast amplification** $\left(P_{\nearrow}^{k,l,m} = P(C_{\text{tst}}^{k,l,m})(1 - P(C_{\text{ref}}^{k,l,m})\right)$.

The visible differences between video sequences convey more information than the other two types of distortions, but especially if the input video pair has different dynamic ranges, the probability map is quickly saturated by the contrast difference that is not necessarily perceived as a distortion. In this case, contrast loss and amplification are useful, which predict the probability of a detail visible in the reference becoming invisible in the test video, and vice versa (Fig. 4). Detection probabilities of each type of distortions are then combined using a standard probability summation function:

$$\hat{P}_{\Delta|\searrow|\nearrow} = 1 - \prod_{k=1}^{K}\prod_{l=1}^{L}\prod_{m=1}^{M}\left(1 - P_{\Delta|\searrow|\nearrow}^{k,l,m}\right). \tag{5}$$

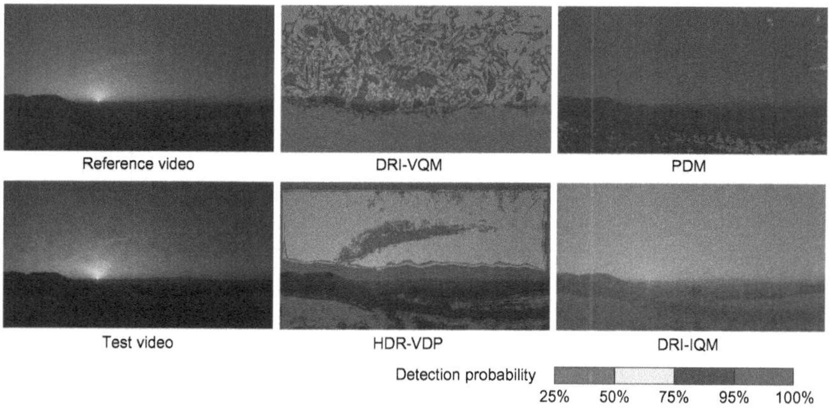

Fig. 4 Comparison of DRI-VQM with other video and image quality metrics, which fail to predict the visibility of the noise pattern present in the test video.

The resulting three *distortion maps* $\hat{P}$ are visualized separately, using an in-context distortion map approach where detection probabilities are shown in color over a low contrast grayscale version of the test video (Fig. 5).

The implementation of DRI-VQM video metric is publicly available online (http://metrics.mpi-inf.mpg.de/) along with other metrics.

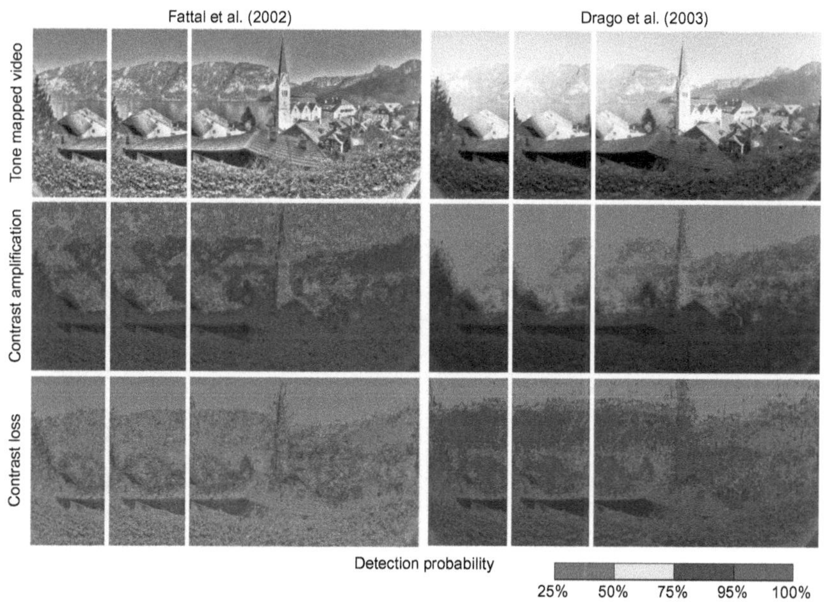

Fig. 5 In-context visualization of the contrast loss and amplification of SDR videos obtained by Fattal and Drago tone mapping operators, with respect to the reference HDR video.

4. HDR-VQM

HDR-VQM, an alternative to DRI-VQM described in Section 3, has been proposed recently [40]. An overview of the HDR-VQM metric is shown in Fig. 6. Similarly to DRI-VQM, the HDR-VQM is a full-reference HDR VQM, hence their building blocks are similar as well. The method is based on signal preprocessing, transformation, frequency based decomposition, and subsequent spatiotemporal pooling, as described in more detail below. However, the main difference resides in the application area. HDR-VQM targets the signal processing, video transmission and related fields, where the distortion of the signal is often considerable and the information about the *overall video* quality is thus, an expected and sufficient measure. Accordingly, HDR-VQM aims to predict human perception of the *suprathreshold* video distortions, which are then pooled to a single number, a measure of an overall video quality.

4.1 Transformation Into Emitted Luminance

First, the input videos are transformed into the luminance values emitted by the display device. This is a difficult problem, because the HDR values encoded in the HDR video are often not calibrated (i.e., relative), and thus, they are merely proportional to the input luminance. Moreover, the accurate display processing model is usually unknown. Instead, the authors of HDR-VQM adopt a simple approximation using a scaling factor as follows. The input HDR videos are normalized by the maximum of the mean of top 5% HDR values of all the frames in the video sequence. A clipping function is finally applied to mimic the physical limitations of the display. This way, the values of the emitted luminance E fit in the

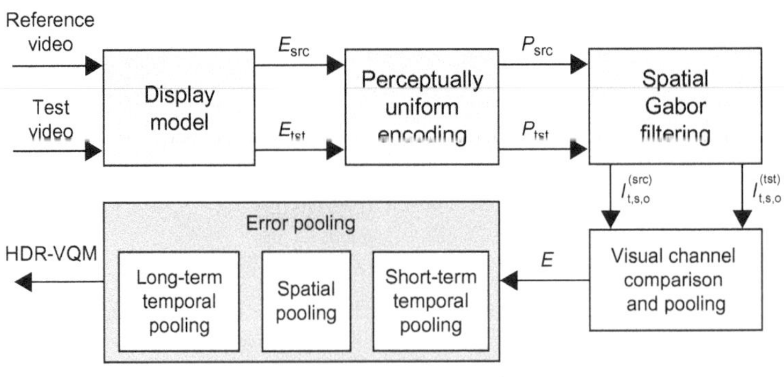

Fig. 6 Data flow diagram of HDR-VQM. See text for details.

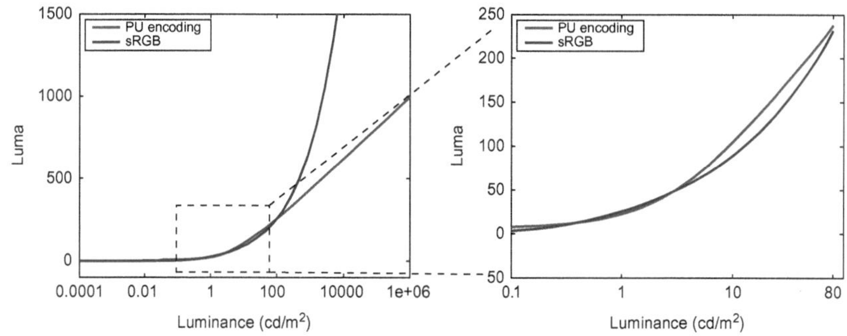

Fig. 7 Perceptually uniform (PU) encoding is backward-compatible with the sRGB nonlinearity. The curve is shown along the entire dynamic range (*left*), and only within the operating range of sRGB (*right*).

range given by the black point of the display and the highest displayable luminance. The values outside this range are saturated, representing the information loss due to the display device.

4.2 From Emitted to Perceived Luminance

The second step approximates the human perception P of the emitted luminance E, which is known to be nonlinear [5], approximately logarithmic. To model this behavior in HDR-VQM, the perceptually uniform (PU) encoding proposed by Aydın et al. [41] was adopted (see Fig. 7). The central idea of the PU encoding is to make differentials of the curve proportional to the luminance detection thresholds. The PU encoding is expected to model the HVS in a better way than a simple logarithmic function, yet still it is only a crude approximation of the HVS luminance response. However, the PU encoding may be implemented efficiently as a look-up table operation (available from http://resources.mpi-inf.mpg.de/hdr/fulldr_extension/).

4.3 Decomposition Into Visual Channels

Similarly to DRI-VQM, the perceived luminance P is subsequently decomposed into visual channels. However, the implemented decomposition distinguishes only spatial frequencies and orientations, leaving temporal processing to the later pooling stage for efficiency. Consequently, the spatiotemporal contrast sensitivity (CSF) of the human visual system cannot be modeled. More specifically, the employed decomposition is based on log-Gabor filters [42], implemented in the frequency domain. This way, the reference and distorted videos are decomposed into visual channels (subbands) $\left\{l_{t,s,o}\right\}$, where $s = 1, 2, \ldots, N_{\text{scale}}$ is the total number of scales, $o = 1, 2, \ldots, N_{\text{orient}}$ is the total number of orientations, and $t = 1, 2, \ldots, F$ is the number of frames in the sequence.

The error in each channel is then computed per frame using a simple bounded measure as follows:

$$E_{t,s,o} = \frac{2l_{t,s,o}^{(\text{src})} l_{t,s,o}^{(\text{dst})} + k}{(l_{t,s,o}^{(\text{src})})^2 + (l_{t,s,o}^{(\text{dst})})^2 + k},$$

where k is a small constant to avoid division by zero.

4.4 Pooling

First, a simple error pooling across scales and orientations is performed. This neglects contrast sensitivity of the human visual system (CSF), which is essential to model near-threshold sensitivity. Assuming suprathreshold distortions, the pooling boils down to a simple equal weighting as follows: $E_t = \frac{1}{N_{\text{scale}} N_{\text{orient}}} \sum_{s=1}^{N_{\text{scale}}} \sum_{o=1}^{N_{\text{orient}}} E_{t,s,o}$, where E_t is a per-frame distortion map. Please notice that no temporal processing has been involved so far, therefore the resulting distortion video $E = \{E_t\}_{t=1}^{F}$ is equivalent to computing an *image* quality metric separately for each video frame t. On the other hand, DRI-VQM described above involves spatio*temporal* decomposition followed by the spatio*temporal* CSF filtering. The distortion video produced by DRI-VQM therefore accounts for temporal behavior of the human visual system.

The subsequent spatiotemporal HDR-VQM pooling step is an interesting way of modeling temporal perception, which has not been considered in the previous steps of the algorithm. Motivated by the alternations in visual fixations, the distortion video E is first divided into nonoverlapping short-term tubes (channels) ST defined by a 3D region $x \times y \times z$. The spatial extent of ST regions $(x \times y)$ is given by the viewing distance, the central angle of the visual field in the fovea and the display resolution. The temporal dimension z is defined by average eye fixation time, and it is set to 300–500 ms in HDR-VQM. Consequently, the *short-term temporal pooling* is performed by computing the standard deviation of each ST tube. This results in spatiotemporal subband error frames $\left\{ST_{v,t_s}\right\}_{t_s=1}^{F/z}$, where v represents the spatial coordinates. Finally, the *spatial and long-term temporal pooling* is performed to yield the global video quality score in a simple way as follows. First, the subband error frames ST_{v,t_s} are pooled spatially resulting in a time series of short term quality scores. Finally, in a long-term temporal pooling, the time series are fused to a single number denoting *overall video quality*:

$$\text{HDR-VQM} = \frac{1}{|t_s \in L_\text{p}||v \in L_\text{p}|} \sum_{t_s \in L_\text{p}} \sum_{v \in L_\text{p}} ST_{v,t_s},$$

where L_p denotes the set with lowest $p\%$ values[1], and $|.|$ is a cardinality of the set. The pooling factor p is set to 5%, however according to the authors, varying p between 5% and 50% does not significantly change the prediction accuracy. It should be noted that contrary to DRI-VQM, HDR-VQM does not explicitly model spatiotemporal *masking* effects.

5. DATA-DRIVEN METRICS

Even though knowledge about the human visual system (HVS) is continuously expanding, many unanswered questions and unverified hypotheses still remain. On that account, we are quite far from having an accurate bottom-up model of the HVS. Therefore, in addition to the bottom-up approaches shown above, top-down *data-driven approaches* based on machine learning are starting to emerge. Machine learning techniques have recently gained a lot of popularity and attention in many research areas. For such methods, it is of crucial importance to provide a sufficient amount of training data. Unfortunately, not many usable datasets exhibiting localized distortion maps measured on human subjects are available. Accordingly, the possibilities of data-driven approaches are currently being explored on simpler, *image* quality assessment task.

More specifically, two experiments [43] were performed where observers used a brush-painting interface to directly mark distorted image regions in the presence and absence of a high-quality reference image. The resulting *per-pixel image-quality datasets* enabled a thorough evaluation of existing full-reference metrics and the development of new machine learning-based metrics.

Specifically, the datasets were utilized to develop a *Learning-based Predictor of Localized Distortions (LPLD)* [44]. LPLD is a *full-reference metric* for synthetic images. The key element of the metric is a carefully designed set of features, which generalize over distortion types, image content,

[1] Both short-term spatial and long-term temporal pooling is performed only over the lowest $p\%$ values.

and superposition of multiple distortions in a single image. Additionally, two new datasets to validate this metric were created and made publicly available (http://resources.mpi-inf.mpg.de/hdr/metric/): a continuous range of basic distortions encapsulated in a few images, and the distortion saliency maps captured in the eye tracking experiment. The distortion maps are useful to benchmark existing and future metrics, and associated saliency maps could be used, for instance, in perceptual studies of human visual attention.

Finally, a data-driven *no-reference IQM* for synthetic images called *NoRM* [45] was proposed. NoRM uses a supervised learning algorithm to predict a perceptual distortion map, which measures the probability of noticing the local distortions on the pixel-level. The proposed metric achieves prediction performance comparable to full-reference metrics. Besides the machine learning machinery, the quality of the results of NoRM is owed to rendering-specific features extracted from the depth map and the surface-material information.

6. OUTLOOK AND FUTURE WORK

Despite many years of active research on image and video quality assessment, the developed metrics are often still far from being comparable to human observers. Existing universal metrics are not mature and robust enough to stand up in all scenarios. However, to overcome this issue, one may develop specialized metrics tailored specifically to the particular problem. Recent examples of such metrics include the quality predictor for image completion [46], or similarity measure for illustration style [47]. Furthermore, measuring vaguely defined quantities like interestingness of images [48], or aesthetic and beauty [49, 50] may also be feasible, perhaps thanks to the machine learning algorithms. Finally, the emerging area of multispectral image and video comparison [51] remains currently almost unexplored.

ACKNOWLEDGMENTS

This work was supported by SoMoPro II grant (financial contribution from the EU 7 FP People Programme Marie Curie Actions, REA 291782, and from the South Moravian Region). The content of this chapter does not reflect the official opinion of the European Union. Responsibility for the information and views expressed therein lies entirely with the authors.

REFERENCES

[1] Z. Wang, A.C. Bovik, Modern image quality assessment, Synth. Lect. Image Video Multimedia Process. 2 (1) (2006) 1–156.

[2] H.R. Wu, K.R. Rao, Digital Video Image Quality and Perceptual Coding (Signal Processing and Communications), CRC Press, Inc., Boca Raton, FL, USA, 2005.

[3] R. Mantiuk, K.J. Kim, A.G. Rempel, W. Heidrich, HDR-VDP-2: a calibrated visual metric for visibility and quality predictions in all luminance conditions, ACM Trans. Graph. 40 (2011), Article 40, 14 pp.

[4] Z. Wang, A.C. Bovik, H.R. Sheikh, E.P. Simoncelli, Image quality assessment: from error visibility to structural similarity, IEEE Trans. Image Process. 13 (2004) 600–612.

[5] S.E. Palmer, Vision Science—Photons to Phenomenology, third ed., The MIT Press, Cambridge, 2002.

[6] S. Daly, The visible differences predictor: an algorithm for the assessment of image fidelity, in: Digital Images and Human Vision, MIT Press, Cambridge, MA, 1993, pp. 179–206.

[7] J. Lubin, Vision Models for Target Detection and Recognition, World Scientific, Singapore, 1995, pp. 245–283 (Chapter: A visual discrimination model for imaging system design and evaluation).

[8] J.A. Ferwerda, F. Pellacini, Functional difference predictors (FDPs): measuring meaningful image differences, in: Conference Record of the 37th Asilomar Conference on Signals, Systems and Computers, 2003, vol. 2, 2003, pp. 1388–1392.

[9] C. van den Branden Lambrecht, O. Verscheure, Perceptual quality measure using a spatio-temporal model of the human visual system, in: IS&T/SPIE, 1996.

[10] P. Lindh, C. van den Branden Lambrecht, Efficient spatio-temporal decomposition for perceptual processing of video sequences, in: Proceedings of International Conference on Image Processing ICIP'96, Proc. of IEEE, vol. 3, IEEE, 1996, pp. 331–334.

[11] C.J. van den Branden Lambrecht, D.M. Costantini, G.L. Sicuranza, M. Kunt, Quality assessment of motion rendition in video coding, IEEE Trans. Circ. Syst. Video Tech. 9 (5) (1999) 766–782.

[12] M.A. Masry, S.S. Hemami, A metric for continuous quality evaluation of compressed video with severe distortions, Signal Process. Image Commun. 19 (2) (2004) 133–146.

[13] A.B. Watson, J. Hu, J.F.M. Iii, DVQ: a digital video quality metric based on human vision, J. Electron. Imaging 10 (2001) 20–29.

[14] A.B. Watson, J. Malo, Video quality measures based on the standard spatial observer, in: IEEE ICIP (3), 2002, pp. 41–44.

[15] S. Winkler, A perceptual distortion metric for digital color video, in: Proceedings of the SPIE Conference on Human Vision and Electronic Imaging, Controlling Chaos and Bifurcations in Engineering Systems, vol. 3644, IEEE, 1999, pp. 175–184.

[16] S. Winkler, Digital Video Quality: Vision Models and Metrics, Wiley, 2005, ISBN 0470024046.

[17] K. Seshadrinathan, A.C. Bovik, A structural similarity metric for video based on motion models, in: IEEE International Conference on Acoustics, Speech and Signal Processing, 2007, ICASSP 2007, vol. 1, 2007, pp. I-869–I-872.

[18] Z. Wang, E.P. Simoncelli, Translation insensitive image similarity in complex wavelet domain, in: Proceedings of the IEEE International Conference on Acoustics, Speech, and Signal Processing (ICASSP '05), vol. 2, ISSN 1520-6149, 2005, pp. 573–576.

[19] M.P. Sampat, Z. Wang, S. Gupta, A.C. Bovik, M.K. Markey, Complex wavelet structural similarity: a new image similarity index, IEEE Trans. Image Process. 18 (11) (2009) 2385–2401.

[20] K. Seshadrinathan, A.C. Bovik, Motion tuned spatio-temporal quality assessment of natural videos, IEEE Trans. Image Process. 19 (2) (2010) 335–350.

[21] H.R. Wu, K.R. Rao, Digital Video Image Quality and Perceptual Coding, CRC Press, Boca Raton, FL, 2005.
[22] Z. Wang, A.C. Bovik, Modern Image Quality Assessment, Morgan & Claypool Publishers, New York, 2006.
[23] C. Chen, W. Chen, J.A. Bloom, A universal reference-free blurriness measure, in: SPIE vol. 7867, 2011.
[24] H. Liu, I. Heynderickx, Issues in the design of a no-reference metric for perceived blur, in: SPIE vol. 7867, 2011.
[25] S. Daly, X. Feng, Decontouring: prevention and removal of false contour artifacts, in: Proc. of Human Vision and Electronic Imaging IX, SPIE, vol. 5292, 2004, pp. 130–149.
[26] E.P. Simoncelli, Statistical modeling of photographic images, in: A.C. Bovik (Ed.), Handbook of Image and Video Processing, Academic Press, Inc., New York, 2005, pp. 431–441.
[27] H.R. Sheikh, A.C. Bovik, L. Cormack, No-reference quality assessment using natural scene statistics: JPEG2000, IEEE Trans. Image Process. 14 (11) (2005) 1918–1927.
[28] A.K. Moorthy, A.C. Bovik, A two-step framework for constructing blind image quality indices, IEEE Signal Process. Lett. 17 (5) (2010) 513–516.
[29] M.A. Saad, A.C. Bovik, C. Charrier, A DCT statistics-based blind image quality index, IEEE Signal Process. Lett. 17 (6) (2010) 583–586.
[30] T. Oelbaum, K. Diepold, A reduced reference video quality metric for AVC/H.264, in: 15th European Signal Processing Conference, 2007, pp. 1265–1269.
[31] L. Ma, S. Li, K.N. Ngan, Reduced-reference video quality assessment of compressed video sequences, IEEE Trans. Circ. Syst. Video Technol. 22 (10) (2012) 1441–1456.
[32] M. Martini, B. Villarini, F. Fiorucci, A reduced-reference perceptual image and video quality metric based on edge preservation, EURASIP J. Adv. Signal Process. 2012 (1) (2012) 66.
[33] R. Soundararajan, A.C. Bovik, Video quality assessment by reduced reference spatio-temporal entropic differencing, IEEE Trans. Circ. Syst. Video Tech. 23 (4) (2013) 684–694.
[34] A.A. Webster, C.T. Jones, M.H. Pinson, S.D. Voran, S. Wolf, An objective video quality assessment system based on human perception, in: SPIE Human Vision, Visual Processing, and Digital Display IV, 1993, pp. 15–26.
[35] J.a. Redi, P. Gastaldo, I. Heynderickx, R. Zunino, Color distribution information for the reduced-reference assessment of perceived image quality, IEEE Trans. Circ. Syst. Video Tech. 20 (12) (2010) 1757–1769, ISSN 1051-8215.
[36] T.O. Aydın, R. Mantiuk, K. Myszkowski, H.-P. Seidel, Dynamic range independent image quality assessment, in: Proc. of ACM SIGGRAPH, vol. 27(3), 2008, article 69.
[37] S. Daly, Engineering observations from spatiovelocity and spatiotemporal visual models, in: Proc. of SPIE: Human Vision and Electronic Imaging III, vol. 3299, 1998.
[38] B.A. Wandell, Foundations of Vision, Sinauer Associates, Inc., Sunderland, MA, 1995.
[39] A.B. Watson, The cortex transform: rapid computation of simulated neural images, Comp. Vision Graph. Image Process. 39 (1987) 311–327.
[40] M. Narwaria, M.P.D. Silva, P.L. Callet, HDR-VQM: an objective quality measure for high dynamic range video, Signal Process. Image Commun. 35 (2015) 46–60.
[41] T.O. Aydın, R. Mantiuk, K. Myszkowski, H.-P. Seidel, Extending quality metrics to full luminance range images, in: Proc. of SPIE: Human Vision and Electronic Imaging XIII, vol. 6806, 2008.
[42] D.J. Field, Relations between the statistics of natural images and the response properties of cortical cells, J. Opt. Soc. Am. A 4 (1987) 2379–2394.
[43] M. Čadík, R. Herzog, R. Mantiuk, K. Myszkowski, H.-P. Seidel, New measurements reveal weaknesses of image quality metrics in evaluating graphics artifacts, ACM Trans. Graph. 31 (2012) 1–10 (Proc. of SIGGRAPH Asia).

[44] M. Čadík, R. Herzog, R. Mantiuk, R. Mantiuk, K. Myszkowski, H.-P. Seidel, Learning to predict localized distortions in rendered images, Comput. Graph. Forum 32 (7) (2013) 401–410.
[45] R. Herzog, M. Čadík, T.O. Aydın, K.I. Kim, K. Myszkowski, H.-P. Seidel, NoRM: no-reference image quality metric for realistic image synthesis, Comput. Graph. Forum 31 (2) (2012) 545–554.
[46] J. Kopf, W. Kienzle, S. Drucker, S.B. Kang, Quality prediction for image completion, ACM Trans. Graph. 31 (6) (2012) 131:1–131:8.
[47] E. Garces, A. Agarwala, D. Gutierrez, A. Hertzmann, A similarity measure for illustration style, ACM Trans. Graph. 33 (4) (2014) 93:1–93:9.
[48] M. Gygli, H. Grabner, H. Riemenschneider, F. Nater, L. Van Gool, The interestingness of images, in: The IEEE International Conference on Computer Vision (ICCV), 2013.
[49] L. Marchesotti, N. Murray, F. Perronnin, Discovering beautiful attributes for aesthetic image analysis, Int. J. Comput. Vision 113 (3) (2015) 246–266.
[50] T.O. Aydın, A. Smolic, M. Gross, Automated aesthetic analysis of photographic images, IEEE Trans. Visual. Comput. Graph. 21 (1) (2015) 31–42.
[51] S. Le Moan, P. Urban, Image-difference prediction: from color to spectral, IEEE Trans. Image Process. 23 (5) (2014) 2058–2068.

CHAPTER 6

Practical Backwards Compatible High Dynamic Range Compression

K. Debattista, T. Bashford Rogers, A. Chalmers
University of Warwick, Coventry, United Kingdom

Contents

1. INTRODUCTION

High dynamic range (HDR) imagery offers an alternative to the traditional method of handling image and video content, frequently termed low (or standard) dynamic range (LDR). While LDR suffers from limitations in the capture, storage, and display of content due to pixels that may be over- or underexposed, HDR is capable of capturing, representing, and manipulating the entire range of real world luminance. The advantages of real world luminance are many; primarily, it affords further realism from allowing manipulation of real world data for applications, such as relighting

High Dynamic Range Video
http://dx.doi.org/10.1016/B978-0-12-809477-8.00006-6

of virtual objects, comparisons to real world, advertising, etc. There is a price to pay for the advantages garnered by HDR. HDR content is represented by floating point data, unlike the single byte per color channel used for LDR content. This entails that at HD resolutions of 1920×1080, uncompressed HDR images require 24 MB of data and a video of 1 min at 30 frames per second would require around 42 GB of data. This is clearly impractical and a number of compression methods for images and video have been adopted to improve storage requirements for HDR content.

There are two broad types of HDR compression methods: those that are backwards compatible, typically consisting of two streams where one of the streams can be viewed on legacy LDR encoders, and those that are dedicated to HDR viewers only. The advantage of backwards compatible methods is that they provide an initial uptake for HDR because they do not require specialized displays or software to view the HDR content. On the other hand, dedicated methods can, potentially, make better use of the available bit rate. The backwards compatible methods require the use of a tone mapping operator [1]; a function that converts the luminance of the HDR content into LDR while attempting to preserve a perceptual match with the original representation. There are a wide variety of tone mappers and parameters required for each, and may require a certain amount of expertise to get the best out of them. After tone mapping, the backwards compatible methods will generate some form of residual or ratio image that is stored as a secondary stream to make up for the lost HDR data in the tone mapped content. This work presents an alternative to the traditional method by simplifying the tone mapping and residual aspect. A single solution is provided based on extracting the luminance channel and a base channel and storing these two. The proposed method, termed practical backwards compatible (pBC), proved successful in experiments conducted, is computationally fast, and does not require any detailed knowledge of tone mappers and parameter settings.

The following section presents the related work and Section 3 describes pBC. Section 4 presents results and analysis of this method against other backwards compatible methods, and Section 5 presents conclusions.

2. RELATED WORK

The dynamic range of real-world lighting situations can vary approximately from 10^{-4} to 10^{8} cd/m^2 [2]. Capturing such a large amount of data requires using special data formats such as Radiance RGBE [3], OpenEXR RGBA

[4], and LogLUV [5] supporting up to 96 bits/pixel (bpp), instead of the traditional 24 bpp integer formats, such as JPEG. Although, the mentioned formats are efficient ways to store real-world imaging data, they have a few drawbacks. First, most of these are floating point data formats and second, the higher bpp cannot be handled by existing video encoders. Therefore, for all practical purposes, uncompressed HDR video data stored in any of these formats cannot be used directly. In order to mitigate this issue, a number of HDR still and video compression methods have been proposed. This section provides a brief background on the different approaches to HDR video compression and briefly discusses some of the compression methods.

2.1 HDR Video Compression Methods

In this section, the focus is on HDR video compression methods apart from JPEG HDR, which is a still image method extendable to video. Both the video and still image version are used in the results section. Banterle et al. [1] provides a comprehensive overview of HDR still image, video, and texture compression. As mentioned earlier, HDR video compression methods can be broadly classified into two groups: higher bit-depth single stream compression methods and backwards compatible double stream methods.

Compression methods following the first approach produce a single raw video stream which allocates higher bit-depth, that is, bit-depth >8, typically 10–14 for the luma channel, and 8 bits for the two chroma channels. Using a range of reversible transfer functions to convert 16-bit floating point luminance to n-bit luma where $n \in [10, 14]$, this approach takes advantage of encoders that are able to support up to 14 bits/pixel/channel. Compression methods following this approach include Perception Motivated HDR video encoding [6], which converts real-world luminance values to 11-bit luma, using perceptual encoding [6], Temporally Coherent Luminance to Luma conversion for HDR video proposed [7], which uses a temporally coherent extension of adaptive logLUV encoding [8] to convert real-world luminance to 12-bit luma, and HVS based HDR video compression proposed by Zhang et al. [9] which converts real-world luminance values to 14-bit luma values using a nonlinear approach similar to Lloyd Max quantization [10]. The computation of chroma values in all the abovementioned approaches are similar to chroma encoding mentioned in logLUV [5].

A few drawbacks of these approaches when used in conjunction with state-of-the-art encoders are that the compressed video files cannot be played back using legacy video players due to the lack of higher bit-depth support, the maximum value of the chroma information, although limited to 8 bits, needs to be passed to the encoder as 11, 12, and 14 bits, respectively because encoders expect similar bit-depths for both luma and chroma and that most existing hardware-based encoders and decoders are limited to 8 bits. Therefore, wide-scale adoption of these compression methods in practice seems unlikely in the near future.

The second group consists of compression methods which split the input HDR data into primary and secondary streams, typically allocating 8 bits/pixel/channel to each of the streams. The primary stream, in a backwards-compatible double stream method is typically a tone-mapped LDR stream, which enables it to be played back using any legacy video player. The secondary stream consists of additional information that can be used to reconstruct the HDR frames. Compression methods following this approach include JPEG HDR [11], HDR MPEG [12], and Rate Distortion Optimized HDR video encoding [13].

JPEG HDR extends the widely used 8 bit/pixel/channel JPEG image format which uses subsampled additional information with precorrection and postcorrection techniques in order to reconstruct HDR frames on the decoder. Originally designed for still HDR images, a fairly straightforward modification can lead to an effective backwards-compatible HDR video compression method. The primary stream consists of LDR frames, created using any given tone mapper. The secondary stream consists of frames created by taking the ratio between the luminance of the input HDR frame and the luminance of the corresponding LDR frame such that $RI(x, y) = \frac{L_{\mathrm{hdr}}(x,y)}{L_{\mathrm{ldr}}(x,y)}$. The luminance of both the HDR and LDR frame are computed using the BT/REC. 709 primaries. The ratio frame is subsequently log encoded, discretized ($RI(x, y) \in [0, 255]$) and subsampled in order to minimize storage requirements.

HDR MPEG, much like JPEG HDR, splits the input HDR frame into two streams. The primary LDR stream is created using a given tone mapping operator and passed through encoder and decoder blocks. Subsequently, both the input HDR and decoded LDR frames and passed through color transformation functions in order to transform both HDR and LDR frames into a similar color space. This transformation helps in the creation of a monotonically increasing reconstruction function (RF), which is subsequently used to predict an HDR frame from its LDR counterpart. Finally, the residual luma differences between the input HDR frame and the predicted HDR frame is quantized using a quantization function and stored as the secondary stream. The quantization function (QF) ensures the residual image is always in unsigned integer range. The algorithm also outputs an auxiliary stream, which stores the RF and QF values per frame to be subsequently used during reconstruction.

The rate distortion optimized HDR video compression method follows a method very similar to JPEG-HDR. The primary stream is created using a temporally coherent TMO, proposed by Lee and Kim [14], essentially a temporally coherent extension of the Fattal TMO [15], which also adds the option of managing color saturation of the LDR frames; although as with the other methods, other tone mapping functions could be used. A log-encoded ratio frame is computed such that $RI(x,y) = \log(\frac{L_{\text{hdr}}(x,y)}{L_{\text{ldr}}(x,y)})$. Subsequently, the minima and maxima of each ratio frame is stored as an auxiliary data structure before the frames are scaled, such that $RI(x,y) \in [0,1]$ and filtered using a cross-bilateral filter [16] in order to reduce noise. Finally, the scaled and filtered ratio frames are discretized and passed to the encoder, and the auxiliary data structure is also stored as a separate file. On the decompression side, the primary frames and secondary frames are read back followed by a saturation correction of the primary frames and inverse scaling of the secondary frames, using information stored in the auxiliary data structure. Finally, the primary and secondary frames are multiplied to reconstruct the output HDR frames.

The primary advantage of the backwards compatible approaches is that the video files can be encoded and decoded using any legacy software/hardware based encoders and decoders, which means hardware support is easy to find and uptake can be quicker because part of the content can be successfully played on legacy LDR software and displays.

3. PRACTICAL BACKWARDS COMPATIBLE COMPRESSION

The backwards compatible methods outlined above share the ability of being flexible, particularly as they enable tone mapping selection and parametrization. However, the choice of which tone mappers are best has been shown to be content dependent [17] and a number of studies do not show consensus for a best method [1]. Recent studies have shown that tone mapping may not even be the best method of displaying HDR data, with certain experiments demonstrating ambivalence of subjects for a single exposure, compared to tone mapped content [18, 19]. Furthermore, within a single image, different tone mappers may perform better than others, as the study and the subsequent hybrid tone mapping approach by

Banterle et al. [20] have shown. This work, rather than offering more complex approaches to backward compatibility, presents an alternative based on a practical equation, which converts the HDR content into two separate streams that are then stored and easily retrieved. The advantage of such a method is that it is straightforward to implement, is computationally fast, and produces good results overall for both video and still image compression.

3.1 Encoding Method

Fig. 1 illustrates the general encoding approach used by pBC. This presents a robust, yet simple implementation of the generalized method described above. Two operations are performed initially, both aimed at fulfilling the variables in Eq. (1). Following the extraction of $L(S_{\text{HDR}})$ (luma stream) and the computation of S_{LDR} (backwards compatible stream) from Eq. (1), the two streams are prepared for legacy encoding, whereby any LDR encoder can be used to encode the two streams. pBC is based around computing the backwards compatible layer based on the following equation:

$$S_{\text{LDR}} = \frac{S_{\text{HDR}}}{L(S_{\text{HDR}}) + 1}, \tag{1}$$

where S_{HDR} is the original HDR image/frame and $L()$ is a function that extracts the luminance of the HDR image/frame using any luminance computation equation such as $L(S_{\text{HDR}}) = 0.2126 \times S^{\text{R}}_{\text{HDR}} + 0.7152 \times S^{\text{G}}_{\text{HDR}} + 0.0722 \times S^{\text{B}}_{\text{HDR}}$ where $S^{\text{R}}_{\text{HDR}}$, $S^{\text{G}}_{\text{HDR}}$, and $S^{\text{B}}_{\text{HDR}}$ are, respectively, the red, green, and blue channels of the HDR image/frame. The equation

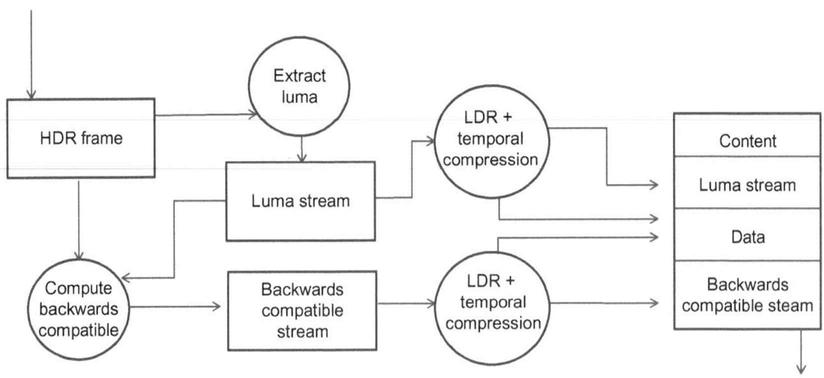

Fig. 1 Encoding method. Pipeline illustrating the encoding method used by pBC.

is applied directly to each pixel. This function is similar to the sigmoid functions used in other methods, but has the unique characteristic of directly computing the color channels through the direct use of the HDR content and its luminance.

The resulting image/frame, S_{LDR} is a full color image/frame that can be encoded with a legacy encoder and that can be viewed on a traditional display. S_{LDR} is encoded as a stream or image and $L(S_{\mathrm{HDR}})$ is also encoded. $L(S_{\mathrm{HDR}})$ can be log encoded to conform with the human visual system's perception of brightness which follows a logarithmic scale or any other encoding mechanism that reduces the range. The resulting $L(S_{\mathrm{HDR}})$ is also encoded using a traditional LDR legacy encoder. In the results presented in Section 4, log encoding of the luminance frame is used.

3.2 Channel Scaling

The dynamic range of S_{LDR} can, on occasion, be larger than 1, a function can be applied to ensure the range is maintained between 0 and 1. In the case of the results presented here, a further computation occurs to the S_{LDR} stream if the values of any of the channels of S_{LDR} exceed 1. This is computed as

$$\mathrm{SC}_{\mathrm{LDR}} = \frac{S_{\mathrm{LDR}}}{(1 + S_{\mathrm{LDR}} \times \mathrm{clamp}((max(S_{\mathrm{LDR}}) * \mathrm{IQR}(L(S_{\mathrm{HDR}}), 0, 1))}, \tag{2}$$

where $\mathrm{SC}_{\mathrm{LDR}}$ is the corrected LDR data, which will not exceed the value of 1. The *clamp*(*value*, *min*, *max*) function is used to clamp results to 1 when required. $\mathrm{IQR}(X)$ computes the interquartile range of the computed HDR luminance channel, and as defined as follows:

$$\mathrm{IQR}(X) = \mathrm{median}(X_R) - \mathrm{median}(X_L), \tag{3}$$

where median() computes the median of a set of data, and X_R is the subset of X, which take values greater than $\mathrm{median}(X)$. X_L is defined analogously. If this step is applied, the corrected data $\mathrm{SC}_{\mathrm{LDR}}$ is stored as the LDR part instead of S_{LDR}.

3.3 Decoding Method

The decoding method is illustrated in Fig. 2 for viewers/displays that support HDR video and Fig. 3 for LDR viewers/displays. The main difference between the two is that when decoding for HDR viewers, the backwards compatible stream is enhanced with the luma stream, while for the LDR viewers the backwards compatible stream is played directly.

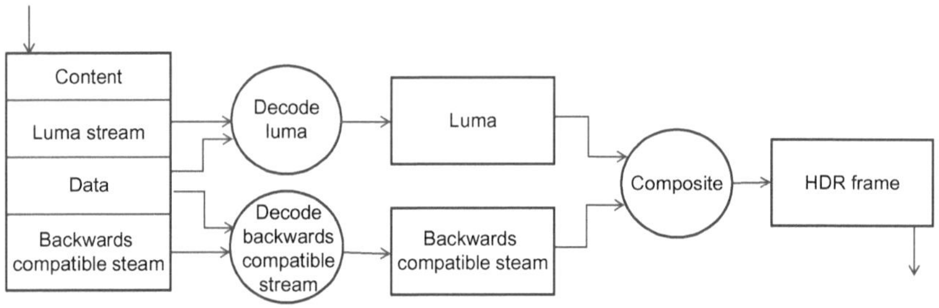

Fig. 2 Decoding method. Pipeline illustrating the decoding method used by pBC for HDR viewing.

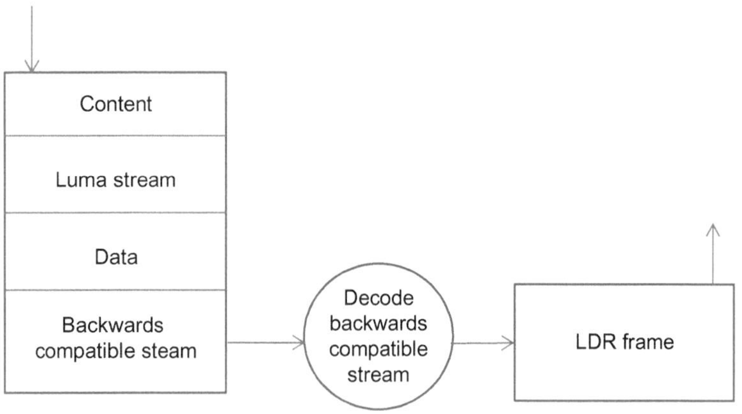

Fig. 3 Legacy decoding method. Pipeline illustrating the decoding method used by pBC for LDR viewing.

When enhancing the backwards compatible stream with the luma stream, Eq. (1) is inverted to reconstruct the HDR stream as $S_{\mathrm{HDR}}^{\mathrm{DEC}} = S_{\mathrm{LDR}} \times (L(S_{\mathrm{HDR}}) + 1)$. If required, Eq. (3) is also inverted and computed prior to inverting Eq. (1). When the luma stream is log encoded (as is the case for the results presented in the next section), then this is log decoded prior to computing any equations. Overall, the decoding process is efficient and easy to implement, making it amiable to hardware implementations.

4. COMPARISON RESULTS

In order to demonstrate the potential of the method, results are demonstrated for HDR video and image compression, comparing the proposed method with other backwards compatible methods. For video compression, all the chosen methods are dual stream methods; furthermore, they all maintain some amount of meta data, which typically results in a few bytes per frame, which for these results are considered negligible. The backwards compatible methods used for testing the results are: pBC (Practical Backwards Compatible) the method being proposed here, the Rate Distortion method [13] (Rate Distortion), the Mantiuk et al. method [21] (MPEG HDR), the Ward and Simmons method [22] originally developed for still images, but that produces good results when extended for video (JPEG HDRv) and, finally, two versions of inverse tone mapping compression methods which use a tone mapper for encoding and an inverse tone mapper for decoding. The inverse methods are presented here as a method of demonstrating the superiority of pBC, when compared to the straightforward inverse tone mapping compression methods to which it may be considered comparable. One inverse method uses the Photographic Tone Mapping Reproduction operator [23] and its inverse [24] (Inverse (Reinhard)), and the second uses a sigmoid similar to the pBC method (Inverse (Sigmoid)) but without the distinguishing quality in Eq. (1) of applying both the **hdr** and *lum*(**hdr**) within the same equation.

Six HDR videos were chosen for computing results. A single frame from each video is shown in Fig. 4; this includes a description of the dynamic range of this content. These videos represent a wide range of HDR possibilities, including CGI (Base) and special effects (Tears). Due to

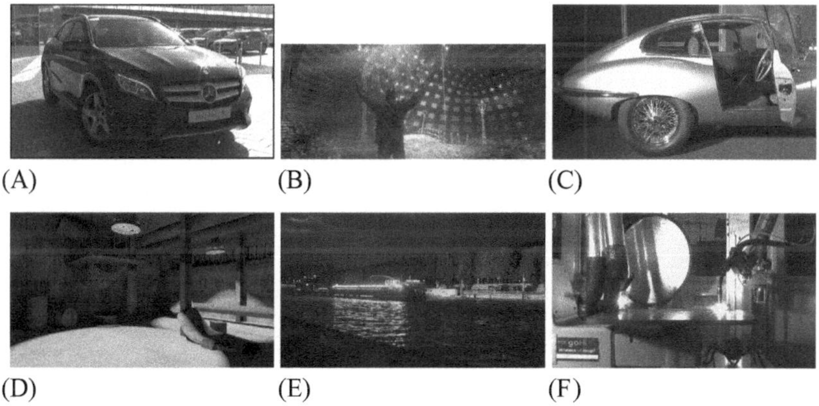

Fig. 4 HDR videos. Scenes used for video results. Dynamic range (DR) calculated as $\log_{10}(\frac{max-min}{min})$. (A) Mercedes; DR = 4.28. (B) Tears; DR = 5.54. (C) Jag; DR = 5.35. (D) Base; DR = 8.37. (E) Seine; DR = 6.30. (F) Weld; DR = 6.48.

the relatively new nature of HDR, there is no single established technique for comparing methods. Three metrics are therefore used to compute the quality of the resulting frames with the original frames, the traditional PSNR method, logPSNR, which attempts to take into account the logarithmic nature of the human visual system when computing differences, and finally, HDR VDP [25] which attempts to recreate the human visual systems response to differences in HDR images.

4.1 Implementation

All the methods were implemented in Matlab using the same framework for most intermediate routines, not specific to an individual method. For each method, two LDR streams are produced and the content of these is explained for each method below. pBC is a direct implementation of the method described in Section 3. The main base stream of pBC contains the backwards compatible stream and the extension stream the luminance. The secondary stream is log encoded to account for the human visual systems response to luminance. JPEG HDRv is a video implementation of the JPEG HDR method [22], as it was seen to produce very good results for video. The method chosen for tone mapping was the Reinhard tone mapper [23], as it has been shown to perform well in psychophysics experiments [17]. The ratio stream is not limited in size, as is the case for still images in JPEG HDR, although the streams produced are typically very small; it is log encoded, as was the case of the extension layer for pBC. HDR MPEG makes use of a temporally-coherent video tone mapper [26], as our implementation using the traditional Photographic Tone Reproduction operator was found to produce poor results for this method. Rate Distortion [13] uses the temporal tone mapper used in the original paper [14]. The bit rate optimization introduced in this method is not employed, as the idea was to compare methods; arguably this optimization could be employed by other methods also. As mentioned above, Inverse (Reinhard) uses the Photographic Tone Mapping Reproduction operator and its inverse. The same sigmoid that inspired pBC is used for Inverse (Sigmoid). The secondary stream for both inverse methods consists of a residual computed from a reconstructed HDR frame's difference from the original.

4.2 Method and Results

The HDR video compression comparison is based around Fig. 5. All HDR frames were encoded with one of the chosen methods, a YUV file was

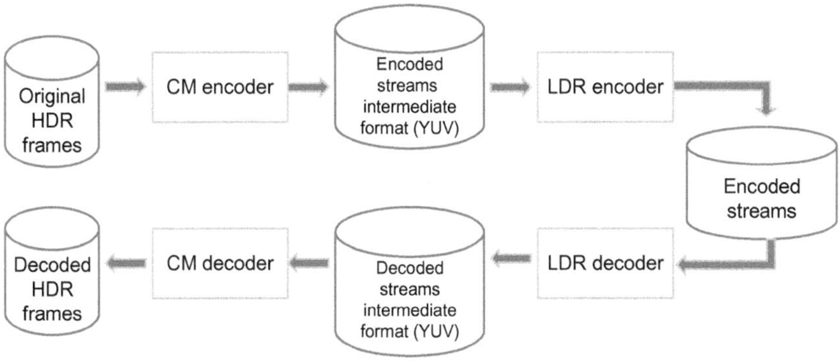

Fig. 5 Method. Results method used for computing results for all compression methods (CMs).

generated and a traditional legacy encoder was used to compress the two YUV files into legacy streams. High Efficiency Video Encoding (HEVC) was chosen as the means of legacy encoding for this set of comparisons, as it represents the upcoming standard for traditional video compression. The x265 application was used to represent HEVC. The settings were set as default maximum quality compression settings (very slow). Each method was encoded with different quantization parameters for each scene with settings of $40, 35, 30, 25, 20, 15, 10, 5, 2, 1$. Once encoded, streams were decoded into YUV streams, which were then converted into streams that were decoded by the various HDR decoding methods, producing a series of HDR frames. The resultant HDR frames were compared with the original frames using the three metrics. Bitrates of the resultant encoded streams were used for computing output bitrates used in the results. The YUV subsampling format of 4:4:4 was chosen with 8-bit depth for both streams. The QP slices were also set equally for both streams. Results are subsequently averaged across all six scenes for each of the metrics.

Fig. 6 shows results for the three metrics. pBC and HDR MPEG perform best overall across all three metrics; pBC performs better than HDR MPEG at PSNR for lower bitrates, and with all metrics at middle bitrates, and HDR MPEG performs best for higher bitrates. JPEG HDRv also performs well, as does Rate Distortion. Rate Distortion would probably perform better with the optimization method included. Both Inverse methods perform relatively poorly. Section 4.4 provides further analysis of the results.

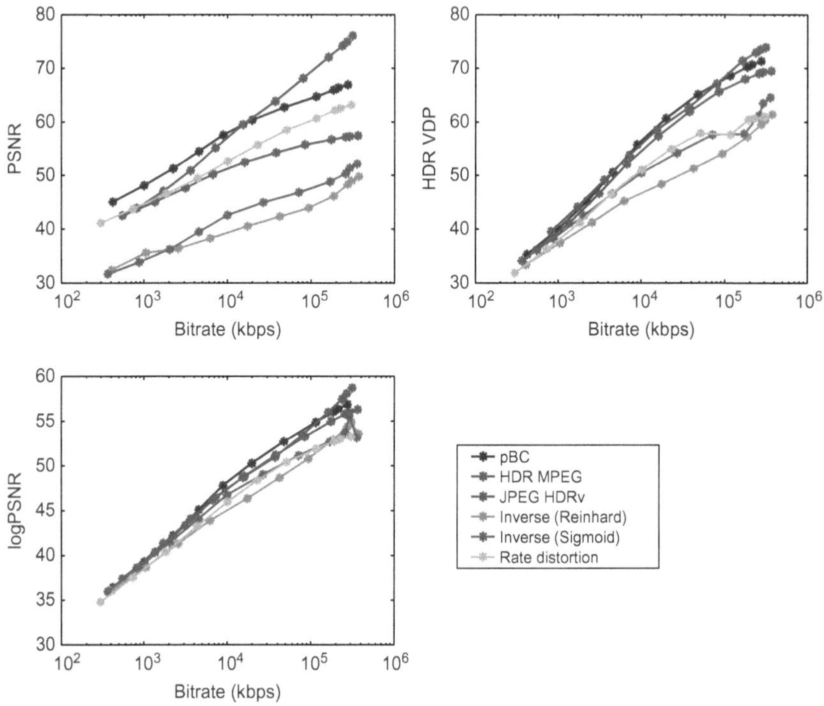

Fig. 6 HEVC HDR video results. Results showing the methods compared for a number of metrics averaged across six scenes.

4.3 Still Images

pBC could also be used for the compression of still HDR images. To demonstrate the approach, the same method used by Ward and Simmons for JPEG HDR [22] is adopted. In this method, a ratio image is encoded in the subband of the JPEG image, and the tone mapped image is stored in the body of the JPEG container. The ratio image is decreased to fit within the subband, typically reduced to <64 KB. pBC was adapted to use the same approach for the results presented in this section. The luminance content of the method is reduced to <64 KB and is stored in the subband. Results are presented for pBC, versus the original JPEG HDR method in Fig. 7 for the 20 images shown if Fig. 8 averaged across all 20 images for the previously used metrics. The images shown in Fig. 7 were all computed with pBC. Bits per pixel (bpp) were computed from the resultant encoded images. Quality parameters were set to 1, 4, 16, 32, 48, 64, 80, 96, and a final no compression value represented as 100. As with JPEG HDRv, the Reinhard

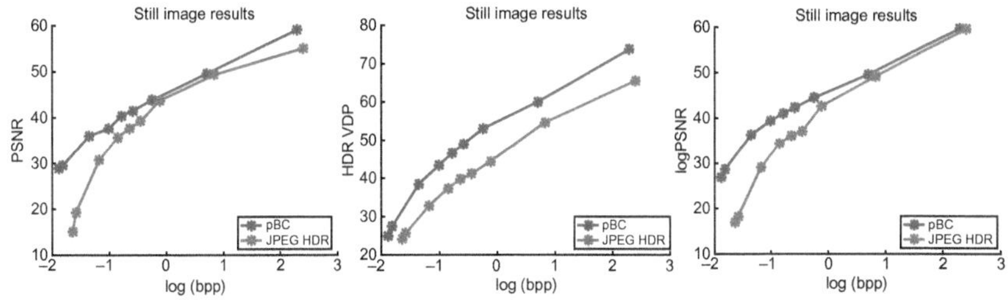

Fig. 7 HDR still image results. Results showing the methods compared for a number of metrics averaged across 20 images.

Fig. 8 HDR images. Images used for still image results computed with pBC.

tone mapper was used for JPEG HDR. Results again demonstrate a good performance for pBC overall.

4.4 Discussion

While details of the methods may vary in terms of implementations, parameters used, tone mappers used, etc., which may modify the results, pBC performs relatively well for a straightforward method, and indicates that there is scope for methods that are unsophisticated. The Inverse methods demonstrate that this straightforward approach is not enough in isolation, and care must be taken with the design of a practical encoding method, which pBC has provided. HDR MPEG, JPEG HDR (and JPEG HDRv), and Rate Distortion, as well as the Inverse methods have the advantage of being able to use different tone mapping operators and therefore, offer more flexibility, but some of this comes at an added cost, potentially in the tone mappers used, for example, local tone mappers may be quite expensive and in the use of complex operators, for example, Rate Distortion makes use of a cross bilateral filter [16]. The Inverse methods lose some of the flexibility, as they require methods that are invertible. Issues like computational complexity play an important role in HDR transmission, which will be a requirement for HDR to succeed in the broadcast domain. With this plethora of options, we believe pBC has an opportunity to succeed as a practical straightforward method that produces good consistent results with little complexities.

5. CONCLUSIONS

This work has presented a straightforward and practical backwards compatible method for HDR encoding of both images and video. As HDR content becomes more popular and is embraced by a larger audience, we suspect that the need for methods that are not too complex to understand and handle become more common place. Our method, pBC, has compared favorably with other similar methods, in terms of results over a number of metrics. While the other, more established techniques have more flexibility as part of their main characteristics, we hope that pBC offers a solution that may be adopted as an alternative to such methods.

ACKNOWLEDGMENTS

Debattista and Chalmers are partially funded by Royal Society Industrial Fellowships.

REFERENCES

[1] F. Banterle, A. Artusi, K. Debattista, A. Chalmers, Advanced High Dynamic Range Imaging: Theory and Practice, CRC Press, Natick, MA, 2011.

[2] E. Reinhard, W. Heidrich, S. Pattanaik, P. Debevec, G. Ward, K. Myszkowski, High Dynamic Range Imaging: Acquisition, Display, and Image-Based Lighting, Morgan Kaufmann, San Francisco, 2010, 650 pp.

[3] G. Ward, The RADIANCE lighting simulation and rendering system, in: Proceedings of the 21st Annual Conference on Computer Graphics and Interactive Techniques—SIGGRAPH'94, ACM Press, New York, 1994, pp. 459–472.

[4] F. Kainz, R. Bogart, Technical introduction to OpenEXR, Tech. Rep., 2009.

[5] G. Ward, LogLuv encoding for full-gamut, high-dynamic range images, J. Graph. Tools 3 (1) (1998) 15–31.

[6] R. Mantiuk, G. Krawczyk, K. Myszkowski, H.-P. Seidel, Perception-motivated high dynamic range video encoding, ACM Trans. Graph. 23 (3) (2004) 733.

[7] J.-U. Garbas, H. Thoma, Temporally coherent luminance-to-luma mapping for high dynamic range video coding with H.264/AVC, in: 2011 IEEE International Conference on Acoustics, Speech and Signal Processing (ICASSP), 2011, pp. 829–832.

[8] A. Motra, H. Thoma, An adaptive Logluv transform for High Dynamic Range video compression, in: 2010 IEEE International Conference on Image Processing, 2010, pp. 2061–2064.

[9] Y. Zhang, E. Reinhard, D. Bull, Perception-based high dynamic range video compression with optimal bit-depth transformation, in: 18th IEEE International Conference on Image Processing (ICIP), 2011, pp. 1321–1324.

[10] S. Lloyd, Least squares quantization in PCM, IEEE Tran. Inform. Theory 28 (2) (1982) 129–137.

[11] G. Ward, JPEG-HDR: a backwards-compatible, high dynamic range extension to JPEG, in: ACM SIGGRAPH 2005 Courses, 2005, p. 8.

[12] R. Mantiuk, A. Efremov, K. Myszkowski, H.-P. Seidel, Backward compatible high dynamic range MPEG video compression, ACM Trans. Graph. 25 (3) (2006) 713–723.

[13] C. Lee, C.S. Kim, Rate-distortion optimized compression of high dynamic range videos, in: Proceedings of the 16th European Signal Processing Conference (EUSIPCO 2008), 2008.

[14] C. Lee, C.-S. Kim, Gradient domain tone mapping of High Dynamic Range videos, in: 2007 IEEE International Conference on Image Processing, IEEE, 2007, pp. 461–464.

[15] R. Fattal, D. Lischinski, M. Werman, Gradient domain high dynamic range compression, ACM Trans. Graph. 21 (3) (2002) 249–256.

[16] E. Eisemann, F. Durand, Flash photography enhancement via intrinsic relighting, ACM Trans. Graph. 23 (2004) 673–678.

[17] P. Ledda, A. Chalmers, T. Troscianko, H. Seetzen, Evaluation of tone mapping operators using a High Dynamic Range display, in: ACM SIGGRAPH 2005 Papers on—SIGGRAPH '05, 2005, pp. 640–648.

[18] A.O. Akyüz, Erik Reinhard, Noise reduction in high dynamic range imaging, J. Visual Commun. Image Represen. 18 (5) (2007) 366–376.

[19] M. Narwaria, M.P.D. Silva, P. Le Callet, R. Pepion, Single exposure vs tone mapped High Dynamic Range images: a study based on quality of experience, in: 2013 Proceedings of the 22nd European Signal Processing Conference (EUSIPCO), IEEE, 2014, pp. 2140–2144.

[20] F. Banterle, A. Artusi, E. Sikudova, T. Bashford-Rogers, P. Ledda, M. Bloj, A. Chalmers, Dynamic range compression by differential zone mapping based on psychophysical experiments, in: Proceedings of the ACM Symposium on Applied Perception, ACM, 2012, pp. 39–46.

[21] R. Mantiuk, A. Efremov, K. Myszkowski, Design and evaluation of backward compatible high dynamic range video compression, Tech. Rep., April 2006.
[22] G. Ward, M. Simmons, Subband encoding of high dynamic range imagery, in: Proceedings of the 1st Symposium on Applied Perception in Graphics and Visualization—APGV '04, ACM Press, New York, 2004, p. 83.
[23] E. Reinhard, M. Stark, P. Shirley, J. Ferwerda, Photographic tone reproduction for digital images, ACM Trans. Graph. 21 (3) (2002) 267–276.
[24] F. Banterle, P. Ledda, K. Debattista, A. Chalmers, Inverse tone mapping, in: Proceedings of the 4th International Conference on Computer Graphics and Interactive Techniques in Australasia and Southeast Asia—GRAPHITE '06, 2006, p. 349.
[25] R. Mantiuk, K.J. Kim, A.G. Rempel, W. Heidrich, HDR-VDP-2: a calibrated visual metric for visibility and quality predictions in all luminance conditions, in: ACM SIGGRAPH 2011 Papers, SIGGRAPH '11, ACM, New York, 2011, pp. 40:1–40:14.
[26] C. Kiser, E. Reinhard, M. Tocci, N. Tocci, Real time automated tone mapping system for HDR video, in: IEEE International Conference on Image Processing, 2012, pp. 2749–2752.

CHAPTER 7

Real-Time HDR Video Processing and Compression Using an FPGA

P. Zemčík, P. Musil, M. Musil
Brno University of Technology, Brno, Czech Republic

Contents

1. HDR VIDEO PROCESSING

High dynamic range (HDR) image and video processing is a trend that addresses the contemporary effort for HDR image, and video capture and reflects the need to acquire and provide highly informational image and video content to the user. As the widely available sensors for HDR images and video require multiple exposures for acquisition of the truly HDR content (the widely available sensors only provide around 12 bits per pixel—12 f-stops—of dynamic range), the processing involves acquisition of multiple images/frames with different exposures, merging of such frames into HDR frames, and storage or further processing of the acquired information. As the flow of data containing the frames of HDR video can be very high, compression of the data is often required and a HDR-specific video compression may have to be involved [1].

This chapter describes a HDR video acquisition and compression system built using mainly programmable hardware that captures multi-exposure HDR video, merges the standard range frames into the HDR frames, compresses the HDR video, and then transmits the compressed

High Dynamic Range Video
http://dx.doi.org/10.1016/B978-0-12-809477-8.00007-8

video over a computer network. This chapter describes the HDR acquisition task in more detail, presents an overall structure of the hardware used for the task, outlines the implementation of the algorithms used for HDR video compression, and shows also the interfaces outline and overall structure of the software. Finally, an overview of the achieved features and exploitation of resources is presented.

2. DESCRIPTION OF THE HDR ACQUISITION AND COMPRESSION TASK

The system described in this chapter is intended for devices, such as surveillance camera units, industrial quality inspection units, or other capture systems with fixed camera and objects of interest under demanding lighting conditions. The anticipated technical characteristics of the system include:

- exploitation of "off the shelf camera," in our case, IO Industries Flare™2KSDI Camera has been selected,
- HDR acquisition of 30 fps HDR video with at least 15 bits per pixel (15 f-stops) and/or 20 fps HDR video with at least 18 bits per pixel dynamic range (the system should be able to switch between the above modes),
- real-time video compression of HDR video using the predefined compression technology developed by goHDR [2], and
- transmission of video over a suitable communication infrastructure, such as Ethernet.

The system as a whole is illustrated in the block diagram in Fig. 1, which also shows the main building blocks of the proposed hardware solution that heavily relies on programmable hardware, namely the FPGA (field programmable gate array) technology [3], specifically the Xilinx ZYNQ architecture that combines the FPGA technology with a couple of ARM CPUs [4]. This system core is accompanied with the H.264

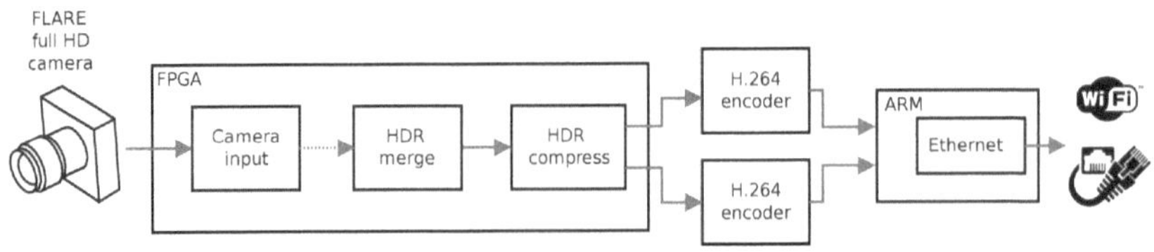

Fig. 1 Proposed HDR acquisition and compression system block diagram.

video compression chips that accomplish the standard task of 8-bit video compression that is a part of the proposed compression scheme [5].

The entire system is completed with several electrical interfaces, such as SDI, and while it can run from batter power, it is also accompanied with battery power supply and charging blocks. The details of individual functional blocks are shown below.

3. ACQUISITION OF HDR VIDEO

The HDR video acquisition process is based on the idea of multi-exposure frame capture [6] and in our case, the input video stream is being formed from 60 fps Full HD video and the desired output can be either 30 fps Full HD with each frame formed from 2 exposures, or 20 fps Full HD video formed from 3 exposures each frame. With sharing the expositions between individual image sets the output can reach up to 60FPS. The video camera used allows for preprogramming in a way so that is generates continuous flow of video frames where the series of (2 or 3) frames are exposed with predetermined exposure times. The image acquisition is typically done in a color format where the color components are simple to handle, and where only the intensity component is processed in HDR processing chain, while the color is preserved independently. In our case, YC_bC_r model is used.

The acquisition process is, in fact, simple from the functional point of view, while it is relatively complex technologically. The algorithm is based on selection of the most suitable pixel for the output frame out of the corresponding pixels from the acquired frames. This is quite a traditional approach [7] and except for some algorithmic difficulties in estimation of the saturation level of individual pixels exposures—especially in case the cameras perform color space conversion, it is not simple to estimate the saturation level—it is simple (see Fig. 2). In our case, we choose the option of selection of only a single exposure, while some approaches tend to perform a linear combination typically of two exposures to avoid transitional effects that did not play important role in our case.

In our approach, the actual representation of pixels in the system is logarithmic and so we used a LUT (LookUp Table) block built from an FPGA memory with static content, connected to the MUX block that encodes the pixels appropriately. The actual content of the LUT intended for calculating the logarithm of the pixels depends on the actual exposure times of the frames and reflects the relations between the individual exposures.

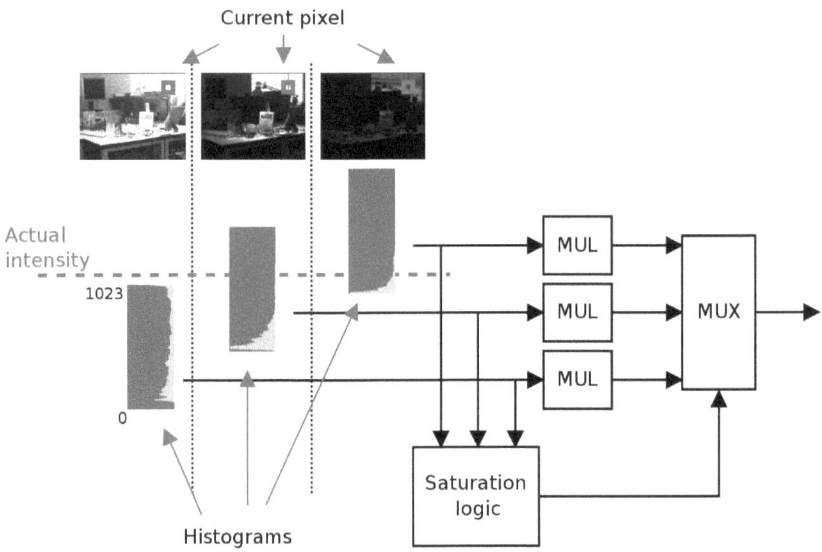

Fig. 2 Selection of the proper exposure level of individual pixels.

Technologically, building of the unit that combines the individual pixels is, however, relatively complex as it involves storage of the 2 or 3 individual frames as they are acquired, and their simultaneous retrieval when the combination needs to be performed—mind that the frames are captures sequentially (see Fig. 3).

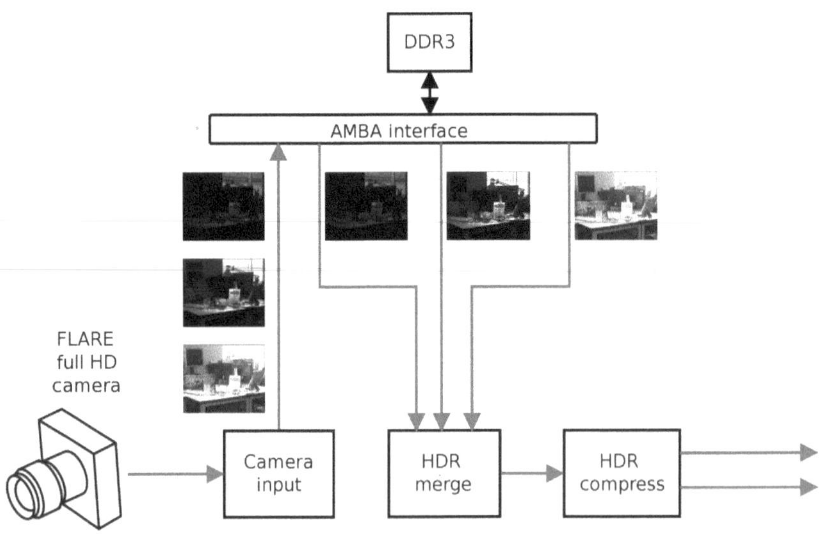

Fig. 3 Actual image acquisition building block.

Moreover, the combining operation (and further processing) must be time overlapped with exposure of further frames of the video, as the flow of data from the camera is continuous and so the combining (and further) operations are performed in parallel. Note, please, that the algorithm does not deal with ghosting effects [8] and deghosting remains subject of future work.

4. HDR VIDEO COMPRESSION IMPLEMENTATION

The actual compression algorithm is performed according to the compression scheme used in [2]. The main idea of the compression is to create a detail-less version of the image, compress its dynamic range, and encode using the standard 8-bit compression (H.264 in our case) into one output stream. Then the differences between the detail-less version of the image and the actual input are encoded as a ratio between the actual pixel value and the corresponding pixel in the detail-less version. Such ratio is encoded in another 8-bit stream along with the color information.

The whole compression process is illustrated in Fig. 4. It contains three blocks in which complexity is high and that are costly to implement—H.264 encoders and Bilateral filter. Two of the blocks, H.264 encoders are, in our approach, implemented in separate electronic components integrating the whole functionality [5]. The third block, however, is not available in this form and had to be implemented in FPGA.

The Bilateral filter block diagram is shown in Fig. 5. The filter is built over image delay lines with built-in register array, which provides access to 11x11 pixel neighborhood at the time. With every pixel at the input, the neighborhood shifts about one position in image. Above the registers, the pipelined computational tree is built, capable of processing one position

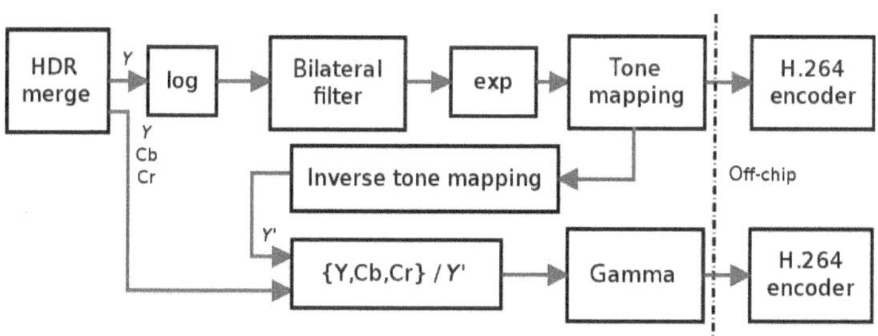

Fig. 4 HDR image compression complete scheme.

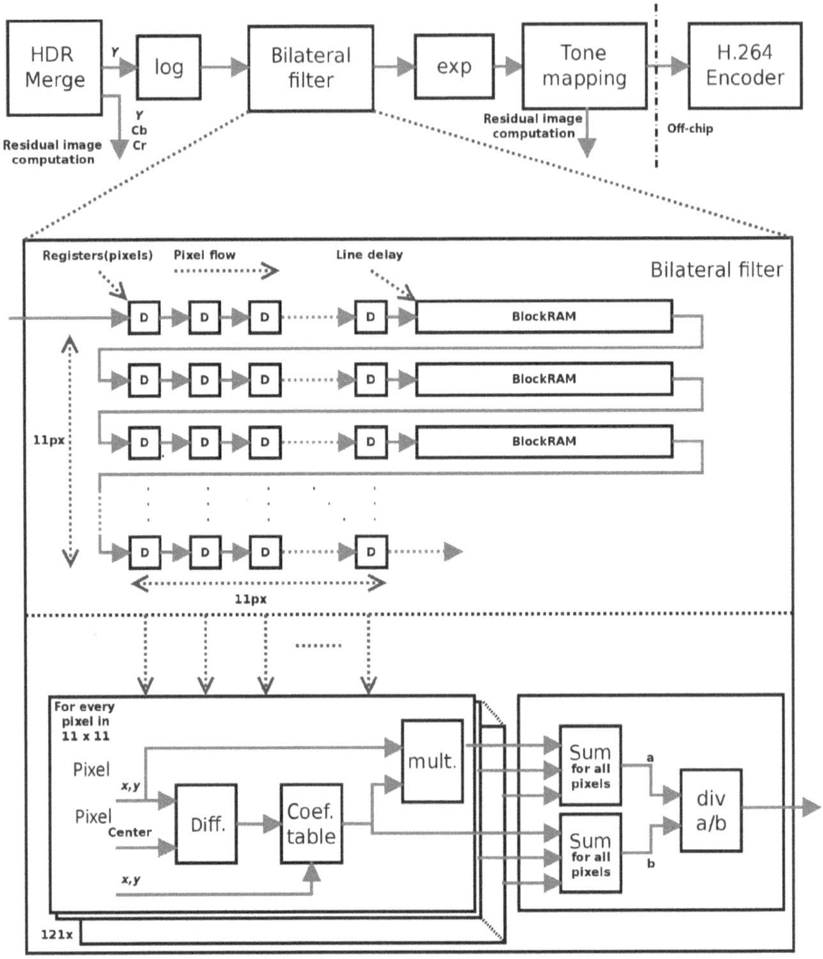

Fig. 5 Bilateral filter block diagram.

in every clock cycle. The computation is performed in the same manner as in the reference implementation of bilateral filter, every pixel gets two weight coefficients, one by distance from the central pixel and the second based on difference of value against the central pixel. All the pixel values are multiplied by their weight and accumulated. At the end, the resultant pixel value is computed by division of this accumulated sum by sum of weight coefficients.

Overall, the design of the acquisition and compression system is done in logarithmic domain. The only difference against reference implementation is using the coefficients composed based on multiples of two, which changes the operation of multiplication into simple addition, which makes the design much easier. The conversion from the linear input is covered in the acquisition block (see the previous paragraph) and the conversion back to the linear scale for the H.264 compression blocks is done using LUTs just before tone-mapping function, which output leads directly to the H.264 encoders.

5. HDR ACQUISITION AND COMPRESSION SYSTEM INTERFACES

Interfacing of the system provides the necessary means of electronic interconnection with the camera on one side and the consumer of the video output on the other side. In our case, the camera input interface was SDI (3 Gb/s), while the output of the system was 1 Gb Ethernet interface at the output of the system. As the H.264 encoders produce their output to an USB interface, this needs to be handled as well. Fortunately, the CPU part of the Xilinx ZYNQ chip is equipped with both Ethernet and USB interfaces, and so these two interfaces were possible to take "as it is." On the other hand, the SDI interface is not available to the CPU part and must be built separately. In fact the SDI to LVDS interface is available as a component and the LVDS interface can be directly handled by the FPGA part of the Xilinx ZYNQ chip (see Fig. 6).

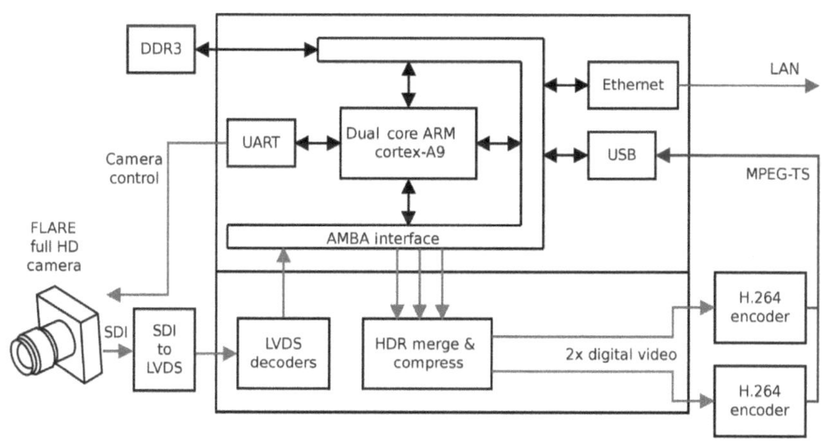

Fig. 6 Figure diagram of the system interfaces.

The interfaces "visible" to the outside world are the SDI input interface and the 1-Gb Ethernet interface. In fact, the data flow over the 1-Gb Ethernet is relatively low, and so it can be converted to wireless transmission over WiFi, or possibly even the LTE mobile network.

6. HDR ACQUISITION AND COMPRESSION SOFTWARE STRUCTURE OVERVIEW

The software involved in the design of the acquisition and compression systems is merely control software, as no important algorithmic part of the acquisition and compression is performed in software, except for the initialization and also configuration. OpenSUSE Linux 13.1 is used as the operating system. The software, more specifically performs the following tasks:

- initialization of the whole system including, the camera, H.264 encoders, and external peripheral devices,
- configuration of the whole system using the web interface through web server run in the CPU part,
- setting the camera parameters and handling continuous change of exposition times,
- video data DMA transfers from SDI into system memory and from the system memory into the FPGA for processing,
- compressed video data transfers from the H.264 encoders into the system memory, and
- output of the compressed video streams over Ethernet interface using the TCP/IP protocol.

7. HDR ACQUISITION AND COMPRESSIONS FEATURES AND RESULTS

Features of the assembled system, from the technical point of view are:

- input of 60 fps FullHD video over SDI interface in YC_bC_r format,
- output of 30 fps FullHD video assembled from 3 exposures (limited by H.264 encoder),
- dimensions $270 \times 180 \times 60\,mm^3$, and
- power consumption of 12 W (without Flare^{TM} 2KSDI Camera).

In designs requiring additional processing of the HDR video, it is important what proportion of resources in FPGA was consumed by the acquisition and

Table 1 FPGA resource consumption

Resources for Zynq XC7Z045	Utilization	Available	Utilization (%)
LUT tables	24700	218600	11
Registers	30405	437200	7
BlockRAM (36 kB)	47	545	9

Fig. 7 Photo of the assembled system.

compression blocks. Also, it is important what power features the system has. These features are shown in Table 1.

A photograph of the assembled system is shown in Fig. 7. Note, please, the "base" FPGA board, the compression modules, and also the SDI interface.

The concept of HDR acquisition and processing will be developed further. The future work includes creation of a compact design that would fit into and be integrated with a camera, possibly including ability to deghost the video. The future work also involves further development of processing algorithms, such as object detection, filtering, and other useful application specific functions.

REFERENCES

[1] Wikipedia, The Free Encyclopedia, High-dynamic-range imaging, Available at: https://en.wikipedia.org/wiki/High-dynamic-range_imaging.

[2] HDR video data compression devices and methods, eP Patent App. EP20,080,012,452 (January 13, 2010), http://www.google.com/patents/EP2144444A1?cl=en.

[3] Field Programmable Gate Array (FPGA), http://www.xilinx.com/training/fpga/fpga-field-programmable-gate-array.htm.

[4] Xilinx Inc., Zynq-7000 All Programmable SoC, Available at: http://www.xilinx.com/products/silicon-devices/soc/zynq-7000.html (accessed June 17, 2016).

[5] Fujitsu, H264 MB86M01/M02/M03 Series Transcoder LSIs, Available at: http://www.fujitsu.com/us/products/devices/semiconductor/h264/mb86m01-2-3.html (accessed June 17, 2016).

[6] K. Myszkowski, R. Mantiuk, G. Krawczyk, High dynamic range video, in: Synthesis Digital Library of Engineering and Computer Science, Morgan & Claypool Publishers, San Rafael, 2008, p. 158.
[7] S. Mann, R.W. Picard, On being 'undigital' with digital cameras: extending dynamic range by combining differently exposed pictures, in: IS&T, 48th Annual Conference, Cambridge, MA, 1995, pp. 422–428.
[8] T.H. Min, R.H. Park, S. Chang, Histogram based ghost removal in high dynamic range images, in: 2009 IEEE International Conference on Multimedia and Expo, New York, 2009, pp. 530–533.

CHAPTER 8

Live HDR Video Broadcast Production

I.G. Olaizola, Á. Martín, J. Gorostegui
Vicomtech-IK4, San Sebastian, Spain

Contents

1. INTRODUCTION

Among the multiple uses of HDR video, live events can get a big benefit from HDR video, especially when it is recorded outdoors with uncontrolled light conditions. HDR technology can provide a better adaptation of cameras to rapidly changing light conditions such as scenes that combine bright sunny areas with dark shadows (football), balls that fly with a very bright sky in the background (golf, football), extremely rapid changes in light conditions (subjective cameras in Formula 1, concerts with flashing lights), etc. All of these cases introduce two main technological challenges: the real-time factor that does not allow any manual intervention nor computationally demanding image data processing step, and the fact that the end-to-end production pipeline has to preserve all the dynamic range information.

High Dynamic Range Video
http://dx.doi.org/10.1016/B978-0-12-809477-8.00008-X

1.1 SDTV, HDTV, and UHDTV

Digital TV and video technologies were originally based on analog video formats. Therefore, standard definition TV (SDTV) takes its visual characteristics from NTSC (480/60i) and PAL/SECAM (575/50i) as it is defined by the ITU Recommendation BT.601 [1] as well as by SMPTE 259M [2]. High definition TV (HDTV) introduces a big change in spacial resolution (720p, 1080i, 1080p) but the color gamut is similar to SDTV and other aspects as the frame rate is not increased. HDTV is defined by the ITU-R BT.709 [3] and shares the same color primaries as sRGB. While the resolution in HDTV can be considered as acceptable for usual distances between the TV screen and the viewer, both, the color gamut (Fig. 1) and dynamic range of the human visual system (HVS) are far beyond the capabilities of HDTV. Therefore, ultra-high definition TV (UHDTV) improves all the dimensions involved in video: resolution, frame rate, color gamut, and dynamic range. UHDTV has been defined by ITU in ITU-R BT.2020 [4]. This recommendation can by summarized as:

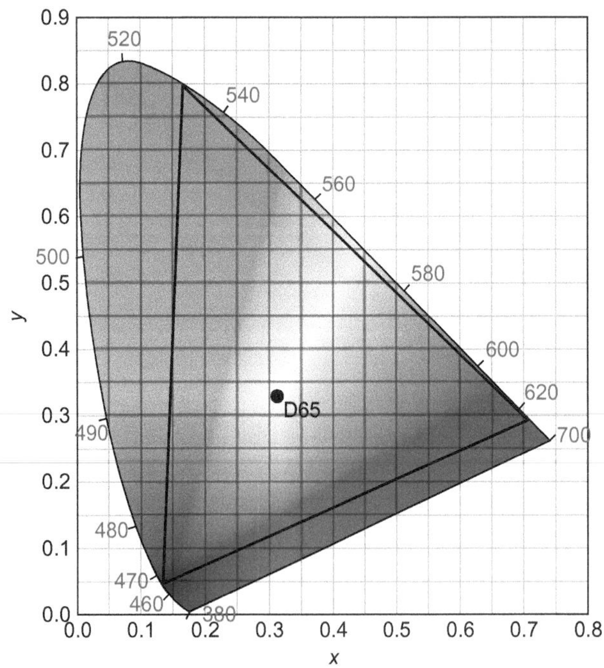

Fig. 1 BT.2020 color space (creative commons BY-SA 3.0).

- Aspect ratio: 16:9 (square pixels)
- Resolution: 4K (3840 × 2160) or 8K (7680 × 4320)
- Frame rate (Hz): 120, 120/1.001, 100, 60, 60/1.001, 50, 30, 30/1.001, 25, 24, 24/1.001
- Only progressive scan mode
- Coding format: 10 or 12 bits per component
- Wide color gamut
- Opto-electronic transfer
- High dynamic range (not defined yet)

Regarding the HDR capabilities of BT.2020, even if there are not specific details about how to process and represent HDR data, there is a real constraint introduced by the 12 bits per component defined as the maximum bit-depth.

1.2 Real-Time HDR Video Processing

Real-time processing of HDR data demands extremely high computational power. A preliminary analysis of the throughput required of raw video is shown in Eq. (1):

$$T_{\mathrm{h}} = R_{xy} \cdot F_{\mathrm{r}} \cdot 3b_{\mathrm{d}} \tag{1}$$

For HD quality, we can consider R_{xy} as 1080i (BT.709 SMTP 292M 1920 × 1080 interlace). In this case at 50fps and 10 bits per channel we obtain 1.45 Gb/s. For 720p, the same equation gives 1.29 Gb/s. In the case of software-based HDR video processing, GPUs are the solution to provide the required real-time performance. For dedicated video processing hardware, the current main limitation comes from the bit-depth that such solutions are able to accept. Typically, hardware solutions do not go beyond 10 bits, a depth that might not suffice for high-quality HDR imaging.

1.3 Dynamic Range Preservation in Live Broadcast Production Workflows

Professional broadcast production workflows are mostly hardware-based. This specific hardware design provides a very good performance together with a high reliability. However, they lack of flexibility to introduce changes in formats or signal characteristics and therefore legacy aspects become a key issue in this kind of infrastructure. Moreover, the big economical investments that these infrastructures require allow only incremental changes that must ensure the complete interoperability with the rest of the elements in the pipeline. In this sense, standards take a key role to ensure this interoperability. Current production systems (HDTV) do not support more than 10 bit flows, which will be reduced to 8 bits when the signal is going to be distributed.

1.4 Main Aspects of Quality Preservation

There are several aspects that become critical in terms of content quality when introducing HDR flows into current broadcasting infrastructures. All these aspects are well known in HD or SD but become even more critical for HDR.

1.4.1 Banding

Banding effects are directly related to quantization errors. As the dynamic range increases, the quantization effects become more noticeable and banding artifacts arise. To avoid this, smaller quantization steps must be taken, but this requires a higher bitdepth (from 8 to 10 or 12 in order to have backwards compatibility with existing infrastructures, and ideally 14 or 16 bits) and nonlinear transform functions based on the HVS that minimize the observable banding effects. Nowadays, this is partially solved by the electro-optical transfer function (EOTF) and opto-electronic transfer function (OETF) mapping curves that will be introduced in a further section.

1.4.2 Chromatic Deviations

Chromatic deviations are side effects of tone mapping operators. Ideally tone mappers only affect the image luminance and chromatic features are preserved as they operate in color spaces where luminance and chrominance values are decorrelated [5]. However, there is always a degree of correlation that can be altered if only the luminance channel is transformed. Boitard et al. [6] evaluated this chromatic deviations on the most popular video tone mapper or transfer functions. According to their study the most robust color space is YD_uD_v derived from CIE $L'u'v'$ (Eq. 2).

$$Y = L, \quad D_u = \frac{u'}{0.62} - 0.5, \quad D_v = \frac{v'}{0.62} - 0.5 \tag{2}$$

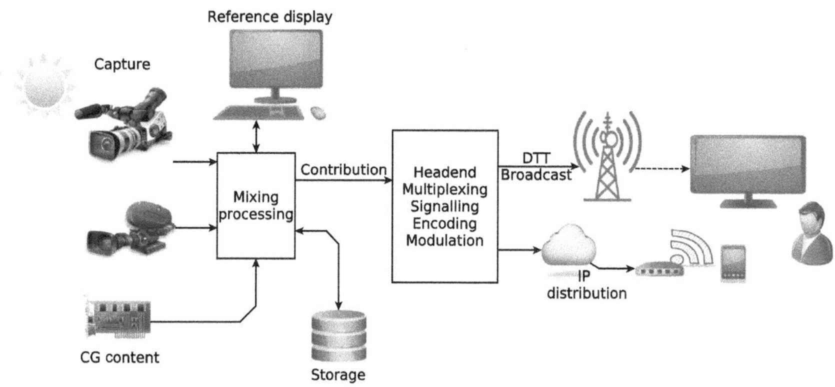

Fig. 2 Basic live TV broadcasting workflow.

2. DIGITAL TV BASICS

Live broadcasting technologies require two main steps, contribution and distribution (Fig. 2).

2.1 Capture

Content can be created either by capturing real scenes with HDR cameras or through synthetic generation by using computer graphics (CG). While CG systems theoretically do not have actual limitations for the dynamic range, cameras are still facing strong technical problems for UHD.

There have been several techniques used to extend the dynamic range of sensors. Bracketing (process of taking several frames at different exposures in order to get a single fused frame with a higher dynamic range) has been extensively used for still images and is used in video capturing as well. However, bracketing introduces ghosting effects over movement and limits the frame rate. The use of two or more cameras at different exposure values combined with bracketing techniques is another solution to obtain a higher dynamic range. Experimental cases have been able to record 18 f-stops by using this approach. However, in this case parallax problems arise. A possible solution to remove the parallax is the use of mirror-based stereo rigs adjusted at 0 parallax. However, all these setups become complex and not feasible for regular TV broadcasts.

The use of multisensor cameras has been the solution adopted by some manufacturers (SpheronVR) that even if useful for other uses, such us engineering or cultural heritage, are too complex and not accurate (blurry) for live broadcasts.

Camera sensors (mostly based on CMOS technology) have rapidly evolved toward very high resolution values reaching 4K. However, this fact implies a smaller sensor size with some negative side effects for dynamic range such as noisier response and smaller dynamic range. Because 4K has become a common feature in most professional camcorders, dynamic range is becoming the next big trend where manufacturers such as Black Magic, Sony, GrassValley, ARRI, etc., can already capture up to 15 f-stops.

This quick evolution of sensors and cameras has introduced some issues that are still being solved by manufacturers. Global shutting (relevant for sports) is still not a feature that all HDR cameras include. Moreover, there is a lack of standard interfaces to get the captured content at full quality (raw or high quality encoding), where manufacturer is mainly applying different OETF curves.

2.2 Manipulation

Manipulation includes applications such as color grading, postproduction, image-based lighting, and scene analysis. While offline processes are done in general by using commercial editing tools such as Avid Media Composer, Da Vinci Resolve from BlackMagic, Speed Grade from Adobe, Sony Vegas, Baselight from FilmLight, Autodesk Lustre, etc. There are still very few tools that allow the real-time processing of HDR video: SGOs Mistika and Vicomtech's Alchemist are two specific solutions for real-time HDR processing. Most of these solutions use GPU-based acceleration to optimize the performance, for example, Blackmagic Da Vinci Resolve can operate with 12-bit RAW content, and apply different techniques for tone mapping.

For software-based video manipulation, OpenCV[1] is probably the most popular image processing tool. The last version at this moment of OpenCV (3.0) includes GPU functions and specific instruction for HDR manipulation, specially tone-mapping operators. Halide[2] shows a novel approach (it was created in 2012) as a functional language that targets different compilers including CUDA[3] and OpenCL.[4]

[1] http://opencv.org/.
[2] http://halide-lang.org/.
[3] http://www.nvidia.com/object/cuda_home_new.html.
[4] https://www.khronos.org/opencl/.

OpenGL[5] and OpenSceneGraph[6] at a higher abstraction level offer advanced GPU-based graphics that can be used for real-time HDR video rendering.

2.3 Contribution

For real-time processes one of the main limitation comes from the physical interfaces. SDI supports up to 12-bit video, and there are commercial solutions able to deal with this bit depth. However, most of the real mixing and processing workflows support operations up to 10 bits and as they are based on specific hardware designs, it is not straight forward to upgrade to 12 bits. This can be a real barrier for the deployment of one of the most relevant use cases for HDR video: live sport transmissions. GPU-based solutions are therefore becoming the main trend to provide this kind of high-end solution with real-time performance. However, there are still no commercial solutions that preserve the whole HDR data without tone-mapping operations along the manipulation processes.

2.4 Distribution

TV signal distribution is currently performed in four main ways. Satellite, cable, or terrestrial broadcast and broadband infrastructures. Classical distribution has been done using broadcast networks that are dedicated infrastructures that require big investments in deployment. Wireless broadcast systems (satellite and terrestrial) do not have scaling limitations as they can provide the content to an unlimited amount or receivers located within the coverage area. On the other hand, cable TV and IP networks can be used for different purposes and are inherently bidirectional and might face problems for massive distribution. The widespread of the Internet availability and the improvement in bandwidth and QoS of IP networks has enabled new business models based on services that are offered directly by Internet Service Providers (in a similar model to cable operators) or even by content providers that do not own any specific infrastructure and operate over the top (OTT). Youtube and Netflix are two of the main OTT providers where Youtube offers prosuming content while Netflix acts as a OTT-based broadcaster. Internet-based delivery networks offer two big advantages compared with classic broadcasting infrastructures. First,

[5]https://www.opengl.org/
[6]http://www.openscenegraph.org/

the continuous evolution and deployment of Internet infrastructures (both wired and wireless) are driven by a huge business model that includes all connected media, services, and devices, ensuring their improvement in a long term. Second, there are not technical or regulatory constraints in content transmission. These two advantages make OTT-based solutions the easiest environment for UHDTV and HDR content distribution. The scalability problem of Internet distribution infrastructures is nowadays solved by Content Delivery Network that basically create intermediate proxies as it can be seen in Fig. 3.

2.4.1 Broadcasting Standards

Broadcasting systems vary between countries, specially for terrestrial transmission. DVB-T [7] (Digital Video Broadcasting) is the most widespread standard, ATSC (Advanced Television Systems Committee) has been adopted in North America and the Japanese System ISDB-T (Integrated Services Digital Broadcasting) [8] is the main standard for South America. DTMB (Digital Terrestrial Multimedia Broadcast) [9] is a system developed and adopted in China and few other countries such as Cuba. Most systems share similar modulation principles (QPSK, OFDM, VSB for ATSC) and include MPEG-2 [10] and MPEG-4 [11] encoding standards. DVB-T is based on COFDM modulation and for HDTV the most adopted solution has been H.264 for video encoding and Advanced Audio Coding [12, 13] for audio. Typical DVB-T parameter configurations establish a bandwidth of around 20 Mb/s in each multiplex and the maximum bandwidth allowed by

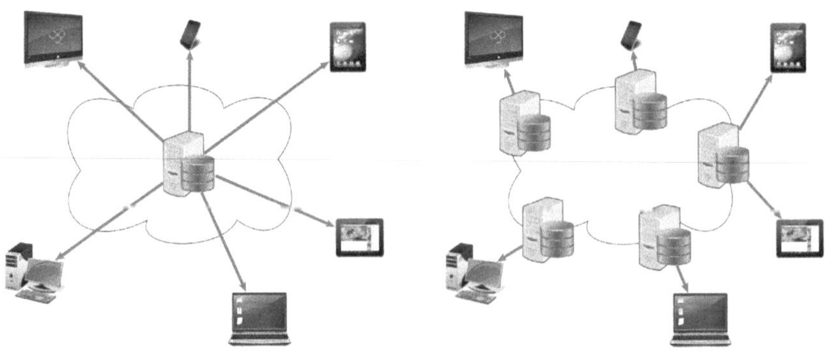

Fig. 3 Single server centralized delivery system vs. CDN architecture.

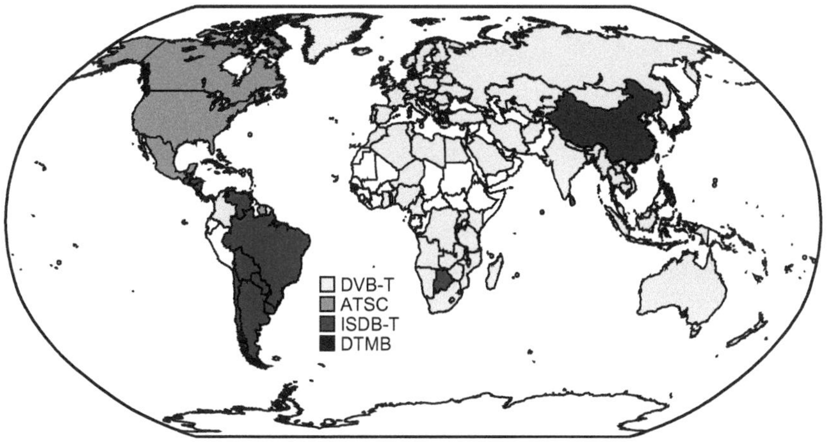

Fig. 4 DTT systems distribution worldwide (public domain license).

the standard is 31.668 Mb/s (64QAM FEC:7/8, Guard Interval: 1/32). The new version of DVB-T, DVT-T2 extends the bandwidth up to 50 Mb/s (Fig. 4).

2.4.2 Encoding

For the highest quality configurations of UHDTV bandwidth requirements can exceed the capabilities of existing distribution infrastructures. 8K@120fps and 12 bits per component means 57.6 times Full HD (1080@50p)[7] and 115.2 times 1080@50i resolution means 16 times more pixels than in Full HD (1080p) and 32 times more than 1080i.

Most proposed solutions for HDR video encoding in broadcast distribution are based on existing LDR video encoders. In order to adapt HDR signals to existing encoders the captured dynamic range has to be quantized to the range where encoding methods can deal with the input signals. As it will be presented in Section 3.1, there are two main approaches to fit HDR data into LDR workflows, the single layer approach where the original signal is mapped through transfer function curves to the desired number of bits and the dual layer approach where the signal is split into two LDR flows.

For distribution, encoding is one of the most critical parts as encoders have to compress the signal in one order of magnitude from contribution

[7] $R = \frac{8K_{\text{res}} \cdot 120_{\text{fps}} \cdot 12_{\text{bpc}}}{HD_{\text{res}} \cdot 50_{\text{fps}} \cdot 8_{\text{bpc}}} = \frac{3840 \cdot 2160 \cdot 120 \cdot 12}{1920 \cdot 1080 \cdot 50 \cdot 8} = 57.6.$

formats, they have to process the signals in real-time and they have to ensure a high level of quality. In the previously proposed workflow and for the sake of backwards compatibility with currently existing broadcast infrastructures, the use of commonly adopted coding standards is proposed. While contribution formats can be used at 12 bits, it is not expected to go beyond 10 bits in distribution. H.264/MPEG-4 AVC [14] is the most extended codec right now and its successor H.264/HEVC [15] is planned to be as one of the main formats for UHD. There are several organizations such as the Blu-ray Disc Association, the High-Definition Multimedia Interface Forum, and the UHD Alliance that have adopted HEVC Main 10 Profile as single layer solution codec [16] where ITU-R BT.2020 wide color gamut is included. This HEVC profile has been defined as HDR10 by the Consumer Technology Association and sets the parameters for the current capabilities of commercial LCD displays (10 bits).

2.5 Reproduction

The rendering and reproduction of HDR content requires displays with two main capabilities. First, the end device has to decode the transmitted signal mapping the information to the dynamic range capabilities of the display that might vary. Second, the display has to be able to reproduce a higher dynamic range than standard commercial devices. To increase the dynamic range, the key aspect of displays is the brightness. While average commercial displays do not go beyond 500 Cd/m^2, professional HDR displays can provide 4000 Cd/m^2 (Dolby Pulsar) or 6000 Cd/m^2 (SIM2[8]). Last consumer electronics devices announced in 2015 (e.g., Samsung and LG displays) claim 1000 Cd/m^2 of peak luminance at 10 bits.

In general, the aforementioned professional devices offer around 14 f-stops which is pretty aligned with the capabilities of current professional video cameras suitable for HDR (e.g., Sony F65, ARRI Alexa, Grass Valley LDX 86, Canon EOS C300, Black-Magic URSA, etc.)

3. INTEGRATION OF HDR FLOWS IN STANDARD BROADCASTING PIPELINES

Broadcast TV workflows encode the different streams (audio, video, metadata) in packets that are signalized according to the corresponding

[8]http://hdr.sim2.it/.

standard (DVB, ATSC, etc., introduces some specific metadata while most content-related metadata are defined by MPEG-2 and MPEG-4).

3.1 Single Layer vs Dual Layer

In order to introduce HDR content to these distribution workflows, signals have to fit in up to 12 bits for contribution and up to 10 bits for distribution purposes. Two address these interoperability issues two main strategies can be followed.

3.1.1 Single Layer

One of them is based on mapping HDR signals into 8, 10, or 12 bits by using the so-called OETF curves that compress the information while trying to keep both the artistic intent and the qualitative perception of the HVS. At the displaying stage, the compressed signal is mapped backed and expanded to HDR by using the EOTF. The whole system can be modeled as the OOTF (Opto-Optical Transfer Function) as it can be seen in Fig. 5.

There are different OETF/EOTF proposals created by Dolby, BBC/NHK, Philips, Technicolor, Sony, ARRI, University of Warwick, etc. Fig. 6 shows the most popular ones, Dolby's PQ published as SMPTE ST 2084 [17], and the Hybrid Log Gamma presented by BBC and NHK, recently standardized as ARIB STD-B67 [18].

4. GPU-BASED REAL-TIME MANIPULATION AND MONITORING

HDR processing algorithms can be typically implemented pixel or regionwise. This fact allows the use of GPUs to boost processes that tend to be computationally intensive. This way, GPUs can provide high performance and much more flexibility than dedicated hardware. Following this approach, Vicomtech-IK4 has developed a prototype (Tebas Platform, Fig. 7) able to perform real-time analysis of 16-bit HDR video flows (histograms, wave forms, vectorscopes, etc.) as well as to manipulate this content for color transformation, mixing, effects, etc. (Fig. 8).

GPU-based solutions use commercial graphic cards that currently are improving faster than CPUs and moreover, offer high flexibility as any new functionality or format interface can be introduced just by adding a

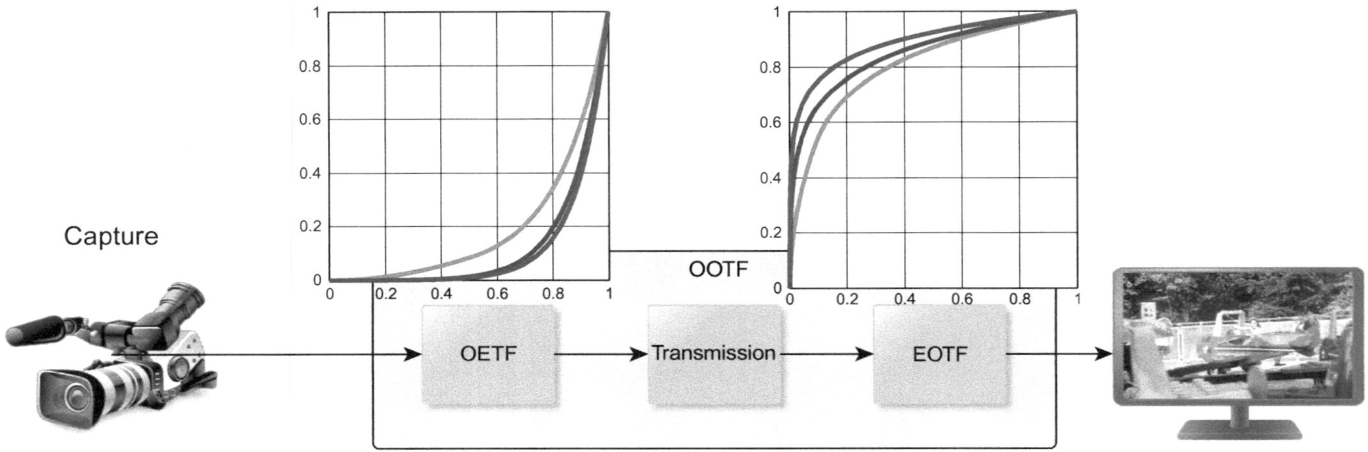

Fig. 5 OOTF workflow. *(HDR picture under creative commons license. Author: Alan Franco.[9])*

[9]https://www.flickr.com/photos/alanfranco/5002543978/in/photolist-8C4m1u-8SRBX8-poK7kR-8C4m3f-8C4kZd-8RD2ua-98acdd-5M2rxE-5M9p4e-8TqtyQ-8C1fjX.

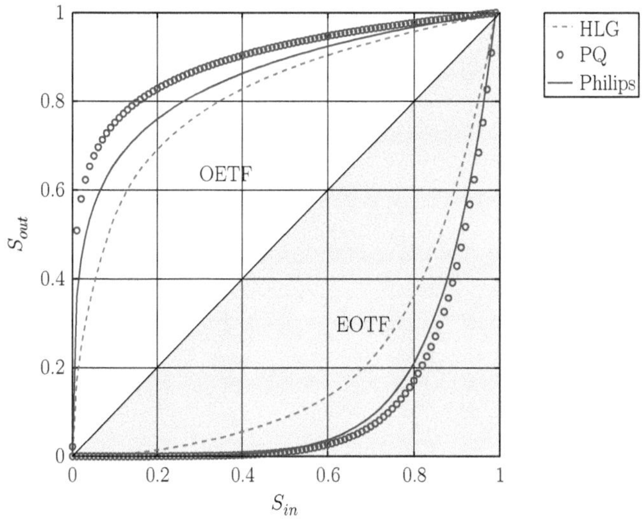

Fig. 6 Main OETF and EOTF proposals.

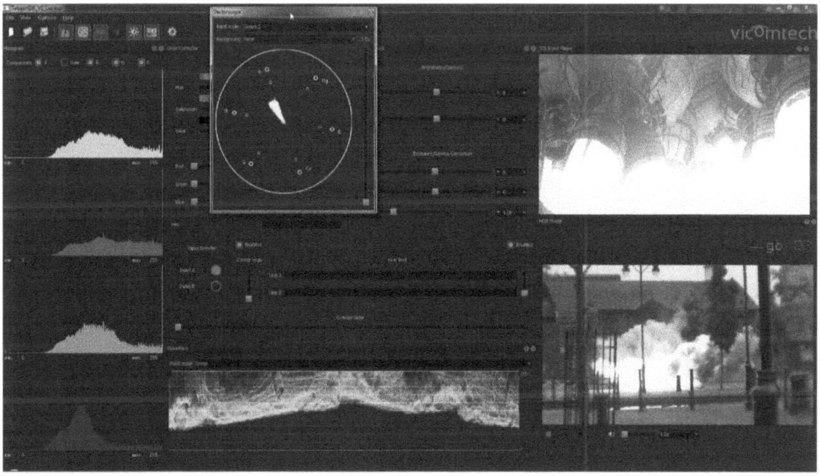

Fig. 7 Screenshot of the GPU-based real-time HDR flow analysis and manipulation platform [Vicomtech-IK4].

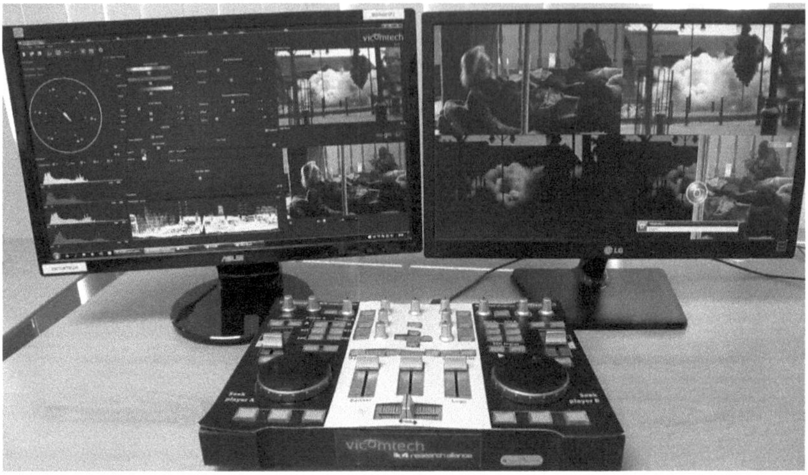

Fig. 8 Operation set-up of the GPU-based real-time HDR flow analysis and manipulation platform [Vicomtech-IK4].

new software module without hardware design constraints. Therefore many video processing functionalities (coding, filtering) are migrating more and more to GPU architectures.

5. CONCLUSIONS

High dynamic range is one of the most relevant characteristics of Ultra HD that introduces a dramatic difference from HD quality and, thus, it can be a key factor to leverage the UHD market. Furthermore, the technological state of the art is ready to provide end-to-end solutions at 14 stops. However, there are interoperability and backwards compatibility aspects that have to be taken into account to ensure a smooth migration from HD infrastructures to HDR workflows. Currently the industry has accepted the use of 12 bits for production and contribution while distribution will be set to 10 bits. Due to this relatively low bitdepth, OETF/EOTF transfer function becomes critical to reduce quantization effects and maximize the quality according to the HVS.

As said, backwards compatibility and cost-effectiveness are two main aspects for the successful deployment of HDR video technology; however, there must be a trade-off with the risk of obsolescence of the adopted solutions. Selected approaches must be future proofing in order to convince the entire value chain (technology manufacturers, content producers, broadcasters, network operators, etc.) to join the HDR video business.

REFERENCES

[1] International Telecommunication Union, BT.601: studio encoding parameters of digital television for standard 4:3 and wide screen 16:9 aspect ratios.

[2] SMPTE ST, For television—serial digital fiber transmission system for SMPTE 259M, SMPTE 344M, SMPTE 292 and SMPTE 424M Signals, SMPTE ST 297:2006 (2006) 1–24, http://dx.doi.org/10.5594/SMPTE.ST297.2006.

[3] International Telecommunication Union, BT.709: parameter values for the HDTV standards for production and International programme exchange, https://www.itu.int/rec/R-REC-BT.709/en.

[4] International Telecommunication Union, BT.2020: parameter values for ultra-high definition television systems for production and International programme exchange, https://www.itu.int/rec/R-REC-BT.2020/en.

[5] M. Pedzisz, Beyond BT. 709, in: Annual Technical Conference & Exhibition, SMPTE 2013, SMPTE, 2013, pp. 1–13.

[6] R. Boitard, R.K. Mantiuk, T. Pouli, Evaluation of color encodings for high dynamic range pixels, in: SPIE/IS&T Electronic Imaging, International Society for Optics and Photonics, 2015, pp. 93941K–93941K.

[7] DVB EBU, ETSI EN 300 744 European Standard (Telecommunications series) Digital Video Broadcasting (DVB); Framing structure, channel coding and modulation for digital terrestrial television.

[8] ARIB, Transmission System for Digital Terrestrial Television Broadcasting ARIB Standard, http://www.arib.or.jp/english/html/overview/doc/6-STD-B31v1_6-E2.pdf.

[9] SAC AQSIQ, GB 20600-2006 Framing Structure, Channel Coding and Modulation for Digital Television Terrestrial Broadcasting System.

[10] ISO/IEC, ISO/IEC 13818-2:2013 Information technology—generic coding of moving pictures and associated audio information—part 2: video, http://www.iso.org/iso/iso_catalogue/catalogue_ics/catalogue_detail_ics.htm?csnumber=61152.

[11] ISO/IEC, ISO/IEC 14496-2, Information technology—coding of audio-visual objects—part 2: visual, 2001.

[12] M. Bosi, K. Brandenburg, S. Quackenbush, L. Fielder, K. Akagiri, H. Fuchs, M. Dietz, J. Herre, G. Davidson, Y. Oikawa, ISO/IEC MPEG-2 advanced audio coding, in: AES-101st 1996, preprint 4382.

[13] ISO/IEC, ISO/IEC 13818-3:1998 Information technology—generic coding of moving pictures and associated audio information—part 3: audio.

[14] MPEG, ISO/IEC 14496-10:2010 Information Technology—coding of audio-visual objects—part 10: advanced video coding, http://www.iso.org/iso/iso_catalogue/catalogue_ics/catalogue_detail_ics.htm?csnumber=56538.

[15] Telecommunication Standardization Sector of ITU, H-265 Series H: Audiovisual and Multimedia Systems Infrastructure of Audiovisual Services—coding of moving video high efficiency video coding, http://www.itu.int/rec/T-REC-H.265-201504-I/en.

[16] Y. Dong, P. Nasiopoulos, M.T. Pourazad, HDR video compression using high efficiency video coding (HEVC), in: UBICOMM 2012: The Sixth International Conference on Mobile Ubiquitous Computing, Systems, Services and Technologies, vol. 6, 2012, p. 7, https://www.google.es/url?sa=t&rct=j&q=&esrc=s&source=web&cd=1&ved=0ahUKEwiA0cCNmdjPAhXGUhQKHQsGDbUQFggeMAA&url=https%3A%2F%2Fwww.thinkmind.org%2Fdownload.php%3Farticleid%3Dubicomm_2012_9_20_10177&usg=AFQjCNFJwNdY-5Jx-LOHqVS8 Qd2QB-HhEA.

[17] ST SMPTE, 2084: 2014: high dynamic range electro-optical transfer function of mastering reference displays, 2014.
[18] ARIB Association of Radio Industries and Businesses, ARIB STD-B67 essential parameter values for the extended image dynamic range television (EIDRTV) system for programme production, http://www.arib.or.jp/english/html/overview/doc/2-STD-B67v1_0.pdf.

PART 3

Display

CHAPTER 9

HDR Video on Small Screen Devices

M. Melo*, M. Bessa*,†, K. Debattista‡, A. Chalmers‡
*INESC TEC, Porto, Portugal
†Universidade de Trás-os-Montes e Alto Douro, Vila Real, Portugal
‡University of Warwick, Coventry, United Kingdom

Contents

1. INTRODUCTION

Nowadays there is a trend toward the widespread use of mobile devices and their evolution allows people to use them for other purposes, rather than simply to communicate. As there is a huge variety of mobile devices, it is important to clarify that in this chapter we refer to mobile devices as all the small screen devices that have both a high level of portability and the capability of reproducing multimedia content. Such devices include smartphones and tablets. Their usage has been increasing at a fast pace and currently it is estimated that there are almost as many mobile-cellular subscriptions in the world as people [1]. According to the last [1] report, by the end of 2015, there will be more than 7 billion mobile cellular subscriptions, corresponding to a penetration rate of 97% globally.

Mobile devices' features promote the consumption of multimedia content, while their high penetration rate is making them a leading platform for the consumption of such content. A recent study by OOYALA [2] gathered data from the online video metrics of an audience of over 200 million unique viewers over 130 countries from 2012 to 2015. This showed

High Dynamic Range Video
http://dx.doi.org/10.1016/B978-0-12-809477-8.00009-1

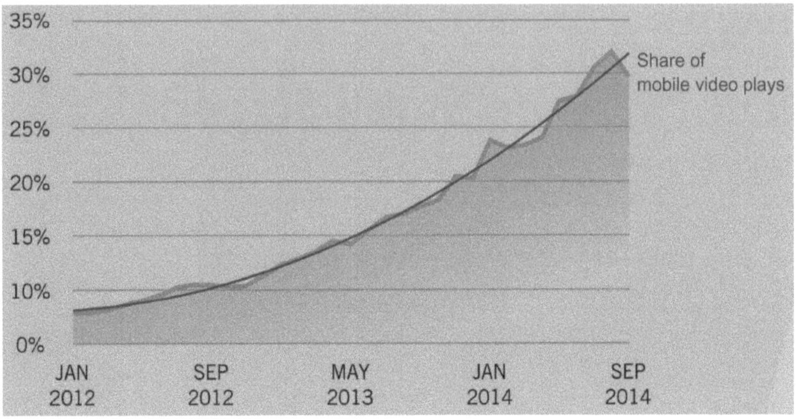

Fig. 1 Mobile devices trend for video consumption. *(Adapted from OOYALA, Global Video Index Q1, OOYALA, USA, 2015.)*

that in the last 3 years, the requests of online video made by mobile devices has increased considerably. While in the first quarter of 2012 the total online video requests made by mobile devices was approximately 5% of all online videos, in the first quarter of 2015, it had grown to around 42% of all online videos (Fig. 1). This corresponds to an increase of over 840% in 3 years and suggests that by the end of 2015, more than half of all online video requests will be made by mobile devices.

In light of such a growing demand, it is expected that the enhanced viewing experience of high dynamic range (HDR) video will be increasingly desired on mobile devices. Although several studies have addressed HDR video visualization, mobile devices are peculiar and present a number of additional challenges when delivering such content on these devices.

2. HDR VIDEO TONE-MAPPING

As there are no HDR displays available on mobile devices yet, it is necessary to reduce the dynamic range of the HDR contents to match the limited dynamic range of the mobile devices' displays. This is achieved through the application of algorithms known as tone-mapping operators (TMOs). So far, only one TMO, the display adaptive TMO [3], has been shown to be successful in a number of evaluations that considered TMOs on mobile devices. In addition, a number of other TMOs have performed well in psychophysical evaluations on mobile devices, including the time-dependent visual adaptation TMO [4], the perceptually based tone mapping of HDR image streams method [5], and the temporal coherence TMO [6].

Display adaptive TMO [3]: Although not specifically designed for mobile, this TMO is able to adapt the tone-mapping process to the display features and usage scenarios through a set of configuration parameters. These include environmental luminance levels, peak luminance, and reflectivity of the display. After proper configuration of the TMO for the specific display, the TMO compares the virtual response of the original HDR frame with the generated frame, using a human vision system (HVS) model and computes additional TMO parameters in order to minimize the visual perceivable differences between the HDR frame and the final low dynamic range (LDR) frame. The configuration flexibility that this method offers is important for mobile devices because their display specifications vary significantly. The method has been seen within a variety of scenarios, ranging from a dark room to outdoors on a sunny day, and the results showed that through the compensation of contrast, it is possible to compensate for any loss that may occur in certain ambient lighting conditions.

Time-dependent visual adaptation TMO [4]: This TMO is inspired by the fact that the HVS does not adapt instantly to big changes in luminance intensities. Thus, in order to deliver a tone-mapped image that is the closest to the real world scene stimulus, this TMO takes into account the appearance change of the scene to match a viewer's visual responses. This is achieved in three steps. The first consists of converting the chromatic values of the image into luminance values for cones and rods in order to determine their retinal response. The second step models the time dependency by using exponential filters to calculate the dynamic response that simulates the process of depletion and regeneration rates. The last step applies a visual appearance model in order to compute the luminance and color appearance models to achieve a match for the viewing conditions.

Perceptually based tone mapping [5]: This TMO is inspired by the tone-mapping approach of Larson et al. [7] and the adaptation over time work of Pattanaik et al. [4]. The method of Larson et al. [7] is used to reproduce visibility of a scene as one would see it in the real scene, while maintaining the overall brightness contrast, and color in the image. One adaptation to this method replaced the original TVI (XXX) model by a new TVIA (threshold vs. intensity and adaptation) model based on the Naka and Rushton work [8]. For the time-adaptation, the TMO uses the approach of Pattanaik et al. [4], but with a different response curve. Small differentials are used along the curves to derive thresholds for the histogram adjustment process, instead of using these directly to compute appearance.

Temporal coherence [6]: This is a postprocessing method that is used in conjunction with a predetermined TMO. Designed specifically of HDR video where temporal coherence needs to be considered, the algorithm comprises two steps: video analysis and postprocessing. Initially the whole video is analyzed to gather information about the global properties from the HDR luminance values of the video. In the second postprocessing step, these data are used to ensure temporal coherence by preserving the relative levels of brightness of the tone mapped video.

2.1 Delivering HDR Video to Mobile Devices

If HDR video is to be successfully delivered to mobile devices, there are a number of unique challenges that need to be overcome. These challenges can be broadly divided into two main groups: hardware related and context related.

2.1.1 Hardware

Hardware plays a key role in HDR video, as it is a demanding technology. Mobile device hardware limitations can compromise the viewing experience. There is a wide spectrum of mobile devices and their features can vary significantly from one model to another. Mobile devices have evolved rapidly in the last few years, and they now possess the computational power equivalent to a conventional PC. However, a number of features that can have direct impact on HDR visualization, such as display features, local storage availability, and power supply [9] remain a problem.

Display Features

The quality of the display varies widely from model to model. This huge variety causes problems when attempting to retarget content for the best viewing experience [10]. Here challenges relate to small size, the available resolution, color gamut, and peak luminance [11].

The first research that investigated the impact of display size on TMO performance was by Urbano et al. [12]. This work only considered static HDR images and compared tone mapping across three different displays: two 17″ displays (one TFT LCD and one CRT) and one 2.8″ display of a smartphone. The experiments considered seven TMOs that were evaluated by a total of 114 participants divided into six groups, randomly assigned by the displays and the scenes. A pairwise comparisons were carried out. The main conclusion of this study was that the participants' preference regarding the tone-mapping performance changes across displays, and that for smaller displays, participants preferred stronger detail reproduction, more saturated colors, and overall brighter image appearance.

Inspired by this work, Melo et al. [13] conducted a series of experiments designed to study the impact of the HDR video tone mapping. This study consisted of a series of psychophysical experiments that were undertaken by a total of 60 participants. Each participant had to rank seven different HDR videos tone mapped by six TMOs. The experiments considered a conventional-sized display of 37″ and a 9.7″ mobile device. The comparisons were made by having the original HDR video displayed on an HDR display as reference. The experiments were divided into two scenarios, one that evaluated the 37″ LDR display, and the other that evaluated the 9.7″ display (Fig. 2).

The author had expected to corroborate the findings of Urbano et al. [12]; that is, that the tone-mapping performance is different across conventional-sized displays and mobile devices. However, interestingly, this

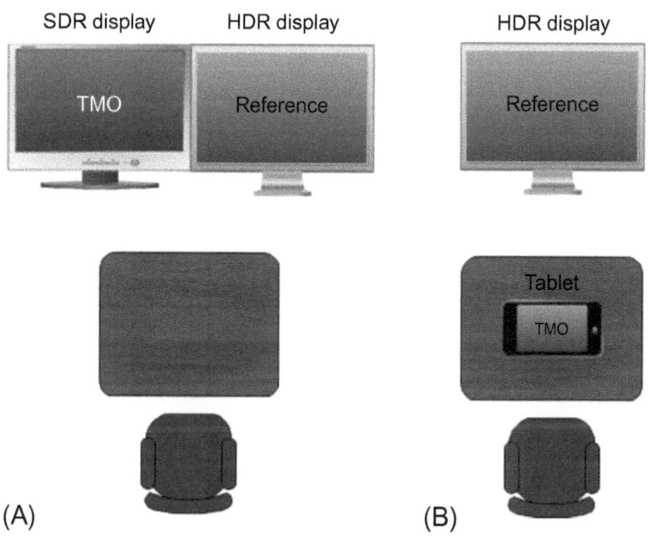

Fig. 2 Experimental setup where (A) represents the scenario where the conventional-sized display was used for the evaluations and (B) the scenario where the mobile device was used. *(Adapted from M. Melo, M. Bessa, K. Debattista, A. Chalmers, Evaluation of HDR video tone mapping for mobile devices, Signal Process. Image Commun. 29 (2) (2014) 247–256.)*

study did not corroborate such findings as the HDR video evaluation showed that participants' preference regarding HDR video tone-mapping performance did not change across displays. The authors explain this result because, with video the differences between displays are not as obvious, as participants did not have much time to focus on the spatial details of the moving images of the video as they could when observing still images.

Local Storage Availability

Mobile devices typically have much less local storage available than conventional PCs. The consumer magazine, *Which?* measured the storage available of eight state-of-the-art smartphones and reached the conclusion that brands announce a certain memory size, but the real memory size available is always considerably lower [14]. For example, in one device for which the manufacturer had announced a memory size of 16 GB, on average, there was only about 11 GB available. This is simply insufficient when one considers that each frame of an uncompressed HDR video sequence at full HD resolution requires 24 MB, and a minute at 30 fps is 42 GB [15]. Although efficient HDR video codecs can help, it is simply not possible to store a collection of HDR videos on a mobile device, as one does with LDR videos.

Power Supply

Portability is a key feature of mobile devices, but the "downside" of this is battery life. When processing high-quality multimedia content on mobile devices, there is a higher battery consumption [16]. Mobile devices' computational power has reached near Moore's law with an almost exponential evolution of technology, but unfortunately this cannot be said for batteries, in which technology has been developing at a much slower pace.

The wide spectrum of mobile devices gives a large range of battery life as shown in Expert Reviews [17] that evaluated the battery life of 60 state-of-the-art mobile devices. The evaluation consisted of playing the Spider-Man 2 movie continuously with the sound output by a set of headphones. The average battery life for the iOS-based smartphones tested was of 13 hours and 25 minutes, followed by the Android-based smartphones with an average battery life of 11 hours and 40 minutes, and Windows-based with an average battery life of 10 hours.

As the average length of a feature film is 1 hour and 30 minutes, the evaluation results show that it is possible to watch a movie without a concern about battery life. There are, however, two facts that one must account for: (1) users do not want to have their battery totally drained by multimedia consumption and compromise the use of the mobile device for communications or something similar and (2) that HDR video has more impact on battery drain than LDR video. Regarding (1) besides the average battery drain caused by the playing the video, users are unaware of power-saving settings on smartphones [18]. Furthermore, they do not respect the charging cycles leading to a significant reduction of battery life [19]. This can be problematic because the multimedia consumption can drain most of the battery, leaving the mobile device unusable for other purposes such as communications. With (2), HDR can aggravate the problem by up to 30% [20], as it requires significant resources due to constant heavy processing. More recent mobile devices are overcoming this possible limitation with more powerful graphics processing units that aim to relieve the CPU load, leading to a more efficient processing and power consumption [16].

2.1.2 Context-Related

Due to their portability, the usage paradigm on mobile devices is different than conventional devices, such as desktop computers or televisions because these are stationary and the environment does not change significantly during usage. The most drastic changes with these stationary displays is when one turns the lights on or off, but even in such events, the environment is controlled and the viewing experience is not compromised for long. With mobile devices, this is different as they are often used "on-the-go" and under varying environments. This can have a negative impact on the viewing experience due to sudden and constant ambient lighting changes or reflections on screen.

Ambient Illuminance

Studies such as Rempel et al. [21] and Melo et al. [22] were conducted to evaluate the impact of the ambient lighting levels on the HDR video visualization. The first study focused on HDR video visualization under different lighting levels, and proved that the ambient lighting levels does have a significant impact on the viewing experience [21] studied the video viewing preferences for HDR displays under varying ambient illumination. Their results show that there are significant differences in preferred display settings under different ambient lighting conditions. This study consisted of two psychophysical experiments that analyzed the potential for visual fatigue and the viewing preferences on next-generation HDR displays. The first was performed by 10 subjects who had to watch 5 movies of approximately 90 minutes each. Five different ambient lighting levels (0.01, 0.75, 8.5, 28, and 74 lux) were evaluated that ranged from a dark ambient lighting that only has the HDR display as the light source, to a poorly lit environment. To avoid subject fatigue due to the length of the video content, the visualization was undertaken in five different sessions. A total of 17 participants did the study. Each subject watched three television episodes of 22 minutes in a single session. Three different ambient lighting (0.01, 70, and 700 lux) were used that ranged from a dark room to a normal indoors environment.

For both studies, participants were distributed randomly across the lighting levels and the video content and placed at a distance of approximately 1.5 m from the display. Participants were encouraged to adjust the brightness and contrast of the display at the start of each session. They were also informed that the brightness and contrast could change spontaneously during the session, in which case they should adjust brightness and contrast to their desired levels. These data were retrieved from a time-stamped log of all brightness and contrast changes made during a session, using a visual fatigue questionnaire, and a more general questionnaire. For this first psychophysical experiment, the relatively small variations in ambient lighting levels did not have an impact on the preferences for brightness or contrast. With regard to visual fatigue, the results showed that participants experienced remarkably little visual fatigue throughout the 90 minutes sessions, as half of participants reported a value of 0 out of 10 and the average level of visual fatigue reported was 1.18 out of 10. The results were uniform across the tested lighting levels. This second study showed that there is a preference toward maximizing the available display contrast. Furthermore, there is a direct relationship between the preferred display brightness and the level of ambient illumination. Participants tended to prefer more brightness for brighter environments

The Rempel et al. study [21] focused on HDR displays and does not consider tone mapping needed for visualizing HDR content on conventional displays. As HDR displays are not yet widespread, it is also important to understand the impact of the ambient lighting level of HDR video tone mapping. Such a study was conducted by Melo et al. [22]. This considered both conventional-sized displays and small screen devices typical of mobile devices. This work extends the previous study on HDR TMO performance across different sized displays [22]. Three distinct ambient lighting environments were considered: dark, medium, and bright. The dark ambient lighting had a measured lighting levels of about 15 cd/m^2, which corresponds essentially to a dark room. The only light was the luminance emitted by the displays used in the experiments. The medium ambient lighting level was about 55 cd/m^2. This roughly corresponds to the average lighting levels of a living room. The bright ambient light scenario had a luminance level of about 1450 cd/m^2, which is similar to indoors on an average cloudy day. Six state-of-the-art TMOs were applied to seven different video sequences. A total of 180 participants were randomly distributed across the different ambient luminance scenarios. They ranked the HDR tone-mapped content on a 37″ LDR display, or on a 9.7″ mobile device display using an HDR display as reference. Fig. 2 shows the experimental setup.

This study corroborated the findings of Melo et al. [22]. One interesting contribution of this work is that results showed that while the TMOs were equally ranked on dark and medium ambient lighting environments, the TMO preference is significantly different when compared within the bright lighting environments. This proves that bright ambient lighting levels have a clear impact on the viewing experience and thus, need to be taken into account. This is also in line with the findings by Rempel et al. [21] that verified differences within the participants' preferred configuration parameters of the display under different ambient lighting levels. Melo et al. also noted that under dark and medium lighting levels, the tone-mapping process should focus more on color and details, while under environments with bright lighting levels, the tone-mapping process should prioritize contrast and naturalness.

One limitation of this study is that the 9.7″ mobile device display is not representative of the wide spectrum of display sizes and features of mobile devices. Smaller display sizes also need to be investigated to see if there is any additional impact on the HDR video tone mapping, as well as to identify any differences between the upper and lower scale of small screen devices as, for example, a tablet and a smartphone.

Reflections

Due to their portability, mobile devices can be easily subject to screen reflections. It is thus important to understand the possible impact of such reflections on the viewing experience. Melo [20] conducted the first

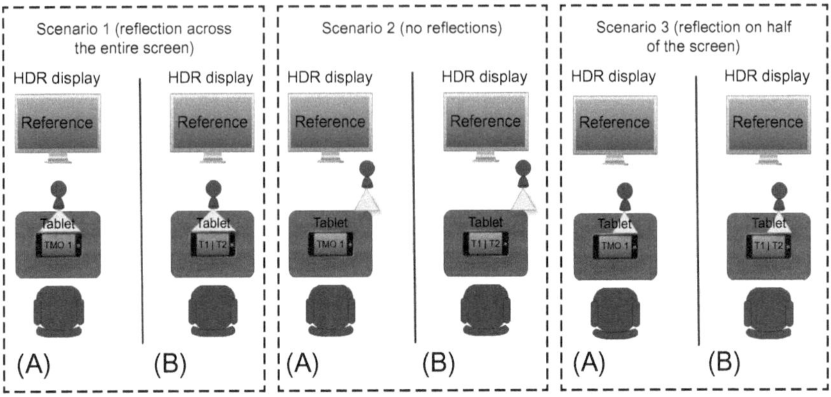

Fig. 3 Experimental setup of the three scenarios studied. *(Adapted from M. Melo, HDR Video for Mobile Devices, PhD, Universidade sde Trás-os-Montes e Alto Douro, 2015.)*

study that investigated tone-mapping performance in situations where a mobile device display is under direct reflections. In total, three experimental scenarios were tested: having the display fully under reflections, only half of the display under reflections, and as a benchmark, without reflections on the display (Fig. 3). As reflections are typical of outdoors scenarios, the authors ensured all the experimental scenarios had a measured ambient lighting level of about 1450 cd/m^2; approximately the same lighting level as a cloudy day.

The experiments considered two state-of-the-art TMOs, as well as a hybrid approach of both in order to understand if such a hybrid could deal with reflections better. A total of 90 participants ranked 6 videos that were all tone mapped with the chosen TMOs. The comparison was made using a reference shown on an HDR display. The results showed that the greater the area exposed to reflections, the larger the negative impact on a TMO's perceptual accuracy. Results also showed that a hybrid approach for tone mapping does not outperform the standard TMOs.

2.2 Available Solutions

The first HDR video player for mobile devices was described by Meira et al. [23]. This targeted two major mobile operating systems: Android and iOS. This player is able to decode HDR videos that were encoded using the goHDR encoder [24]. The player is composed essentially of five modules: Storage, Decoder, HDR Model, Render, Screen (Fig. 4).

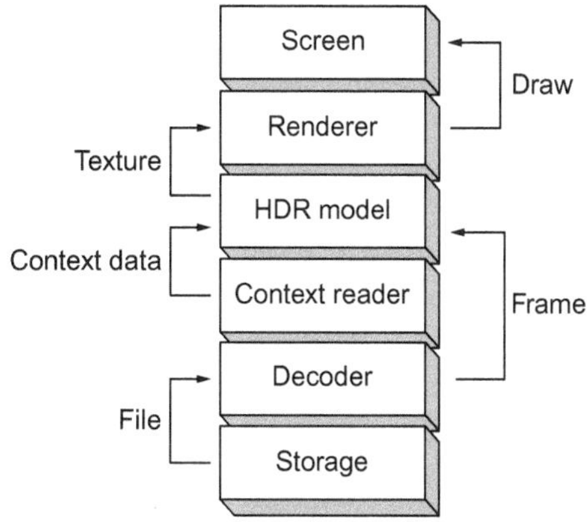

Fig. 4 HDR video player for mobile devices architecture. *(Adapted from C. Meira, M. Melo, M. Bessa, L. Barbosa, A. Chalmers, Performance of HDR video in mobile devices, in: HDRi2013—First International Conference and SME Workshop on HDR Imaging, 2013, pp. 17–21.)*

The module "Storage" is responsible for loading the HDR video content from memory and making it available to the "Decoder." The "Decoder" decodes the files and makes available all the information needed for playing the content. The HDR model is the main module. The module processes all the information that is passed by the decoder, and it applies at the GPU level a specific shader that was specially designed for the purpose.

In order to validate the proposed HDR video player, a performance evaluation was conducted with regard to fps, battery usage, average CPU load, and average RAM usage. The results showed that it is possible to reproduce HDR video on state-of-the-art devices and achieve the desired number of fps and with a satisfactory battery consumption. However, older devices can struggle to keep a satisfactory fps rate.

Melo [20] extended the work of Meira et al. [23] and proposed a solution that supports both state-of-the-art mobile devices and legacy devices. This approach also takes into account the visualization of both offline and online HDR video. The visualization context is exploited to optimize the tone-mapping process and, consequently, the viewing experience. This HDR video delivery system is composed of an HDR video player and an HDR video streaming server (Fig. 5).

State-of-the-art devices are capable of decoding HDR video locally, while legacy devices are unable to do so. The first thing the architecture

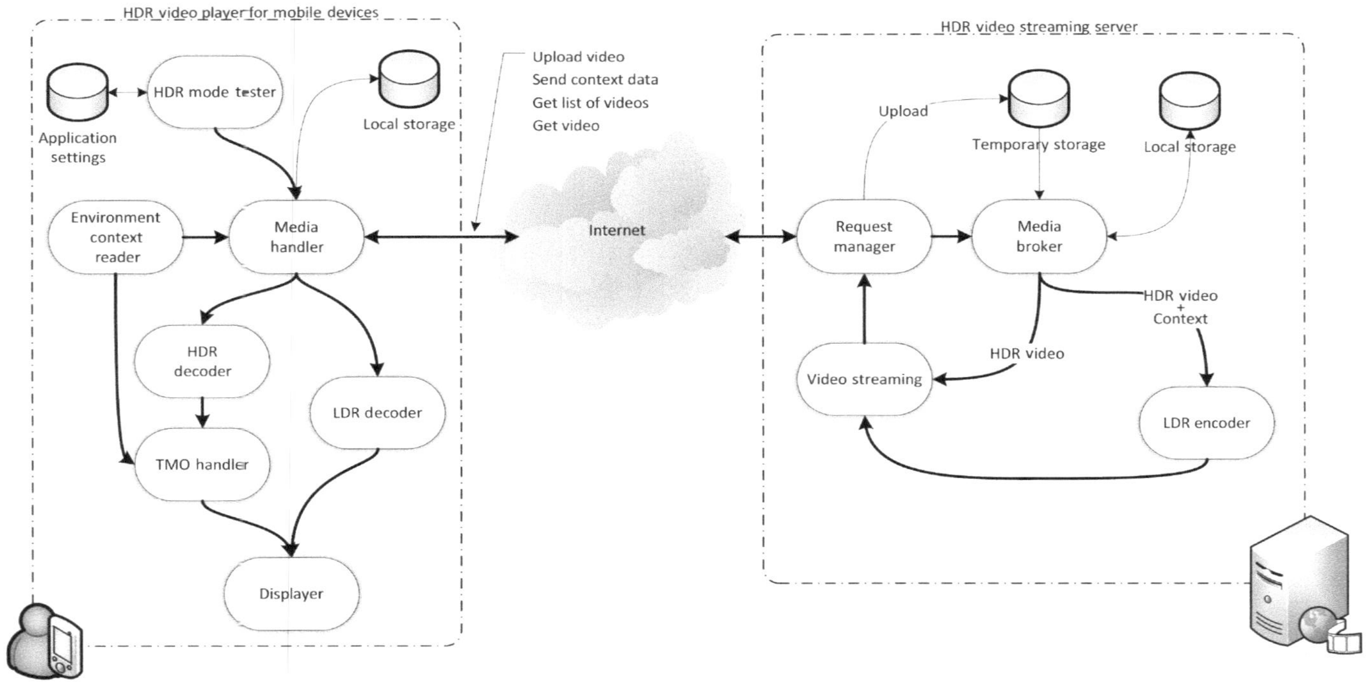

Fig. 5 Architecture of the HDR video delivery system for mobile devices. *(Adapted from M. Melo, M. Bessa, K. Debattista, A. Chalmers, Evaluation of tone-mapping operators for HDR video under different ambient luminance levels, Comput. Graph. Forum, 2015.)*

does, thus is determine the capability of the device by means of the module "HDR Mode Tester." If the mobile device is capable of decoding HDR locally and the video is stored on the device, the "Media Handler" module loads the video from the storage. If the HDR video is located at the HDR video streaming servers, "Media Handler" will communicate with the remote servers in order to request an HDR video stream.

The HDR video or HDR video stream is forwarded to the "HDR Decoder" module which decodes the HDR frames and all the associated information. This is passed to the "TMO Handler" module that gathers context information through the "Environment Context Reader." The frames are then rendered, using all this information. The rendered frames are then forwarded to the "Displayer" for drawing the frames on the display at the encoded frame rate.

When the mobile device does not support HDR video decoding locally, the "Media Handler" will be responsible for uploading the HDR video to the HDR video streaming servers. It also uploads the context information gathered by the "Environment Context Reader" module in order to have the server's transcoding and provide a proper tone-mapped version of the HDR video. It is also possible to browse a list of online HDR videos that is provided by the "Media Handler" and request the proper LDR stream.

The HDR video streaming server is essentially responsible for managing the requests made by the HDR video player and providing it with a proper video stream. There are two possible scenarios: provide an HDR stream or provide a tone-mapped LDR stream. In the first case the process is simple, the "Request Manager" forwards the request to the "Media Broker," that will load the HDR video from the server's storage and make it available to the "Video Streaming" module. This will, in turn, provide an HDR stream to the HDR Video Player. If the request is made by a mobile device that does not support HDR video decoding and the video is located at the device, the "Request Manager" is responsible for processing the HDR video upload and storing it into temporary storage. The "Media Broker" is the module responsible for loading the HDR content from the respective storage and forwarding it to the "LDR Encoder," along with the context data sent by the client application. The "LDR Encoder" module transcodes the HDR video into a proper LDR stream, taking into account the context data that is posteriorly forwarded to the "Video Streaming" module. This will handle the data and provide a proper stream to the HDR Video Application.

This novel HDR video delivery architecture for mobile devices was implemented and evaluated. The evaluation tested a total of six mobile devices (three Android devices and three iOS devices) under the four possible scenarios: local HDR decoding, HDR video streaming, local tone-mapped LDR video decoding, and tone-mapped LDR video streaming. Note that the tone-mapped LDR video could be compared to a conventional video. Two different TMOs were tested. One was best suited for low and medium ambient lighting scenarios, and one that is optimal for bright ambient lighting levels. A total of three videos were tested and each condition was 90 minutes; the average length of a feature film.

The evaluation measured the performance of the HDR video delivery system, as well the impact of the HDR video visualization against the tone-mapped LDR video. The average fps number, the battery drain and the average CPU load were recorded. Overall, it was found that HDR video impact is not much different than LDR video, except for battery consumption. For the tested scenarios, an impact was reported that can reach up to 30% when compared to the reproduction of LDR videos. The paper showed that interactive tone mapping is possible and if the environmental context changes, the HDR video streaming server can adjust the tone-mapping parameters and deliver the appropriate stream to the HDR video player. This should help ensure an optimal visualization experience. Furthermore, the proposed solution supports both state-of-the-art as well as legacy mobile devices. Although the real-time goHDR encoder was used [25], the system is designed to include other HDR video codecs in a seamless manner.

2.3 Future Trends

Moore's law is an observation that is still valid today and that can be applied to mobile devices and their evolution: the power of a device doubles every 2 years. Due to this, associated with the huge demand for mobile devices, it is expected that mobile device features will continue to develop at a rapid pace. This continual development is likely to mean that HDR video can be played on future mobile devices without any of the current hardware constraints. What is not clear is when HDR displays are likely to appear on mobile devices. Such displays have significant power and heating challenges. Until this happens, tone mappers will be required that are best able to deliver an enhanced viewing experience no matter where, or when, the HDR video content is being viewed. These will need to be dynamic and adapt as the context changes. Furthermore, new sensors may appear, such as sensors able to detect reflections. These can be exploited by the TMOs; possibly even using different TMOs, or parameters when tone mapping those parts of the image under the reflection and those which are not.

In addition to the devices themselves, connectivity and network access will continue to improve. This will allow the live streaming of HDR content without constraints, providing a step change in imaging for a wide range of applications, such as surveillance, entertainment, medical imaging, etc.

ACKNOWLEDGMENTS

This work was supported by the project "TEC4Growth – Pervasive Intelligence, Enhancers and Proofs of Concept with Industrial Impact/NORTE-01-0145-FEDER-000020" is financed by the North Portugal Regional Operational Programme (NORTE 2020), under the PORTUGAL 2020 Partnership Agreement, and through the European Regional Development Fund (ERDF).

REFERENCES

[1] International Telecommunication Union, The world in 2015: ICT facts and figures, International Telecommunication Union, Geneva, Switzerland, 2015, http://www.itu.int/en/ITU-D/Statistics/Documents/facts/ICTFactsFigures2015.pdf (accessed October 7, 2015).
[2] OOYALA, Global Video Index Q1, OOYALA, USA, 2015.
[3] R. Mantiuk, S. Daly, L. Kerofsky, Display adaptive tone mapping, TOG 27 (3) (2008) 1–10.
[4] S. Pattanaik, J. Tumblin, H. Yee, D. Greenberg, Time-dependent visual adaptation for fast realistic image display, in: Proceedings of the 27th Annual Conference on Computer Graphics and Interactive Techniques (SIGGRAPH'00), 2000.
[5] P. Irawan, J. Ferwerda, S. Marschner, Perceptually based tone mapping of high dynamic range image streams, in: K. Bala, P. Dutré (Eds.), Sixteenth Eurographics Conference on Rendering Techniques (EGSR'05), Aire-la-Ville, Switzerland, 2005, pp. 231–242.
[6] R. Boitard, K. Bouatouch, R. Cozot, D. Thoreau, A. Gruson, Temporal coherency for video tone mapping, in: Applications of Digital Image Processing XXXV, vol. 8499, 2012.
[7] G. Larson, H. Rushmeier, C. Piatko, A visibility matching tone reproduction operator for high dynamic range scenes, IEEE Trans. Visual. Comput. Graph. 3 (4) (1997) 291–306.
[8] K. Naka, W. Rushton, S-potentials from luminosity units in the retina of fish (Cyprinidae), J. Physiol. 185 (3) (1966) 587–599.
[9] R. Bilcu, A. Burian, A. Knuutila, M. Vehvilainen, High dynamic range imaging on mobile devices, in: 15th IEEE International Conference on Electronics, Circuits and Systems, 2008 (ICECS 2008), St. Julien's, 2008, pp. 1312–1315.
[10] J. Ross, R. Simpson, B. Tomlinson, Media richness, interactivity and retargeting to mobile devices: a survey, Int. J. Arts Technol. 4 (4) (2011) 442–459.
[11] L. Chittaro, Visualizing information on mobile devices, Comput. 39 (3) (2006) 40–45.

[12] C. Urbano, L. Magalhães, J. Moura, M. Bessa, A. Marcos, A. Chalmers, Tone mapping operators on small screen devices: an evaluation study, Comput. Graph. Forum 29 (8) (2010) 2469–2478.

[13] M. Melo, M. Bessa, K. Debattista, A. Chalmers, Evaluation of HDR video tone mapping for mobile devices, Signal Process. Image Commun. 29 (2) (2014) 247–256.

[14] Which? Tech Daily, Phone storage compared—Samsung S4 still in last place, 2014, http://blogs.which.co.uk/technology/phones-3/phone-storage-compared-samsung-s4-still-in-last-place/ (accessed October 7, 2015).

[15] F. Banterle, A. Artusi, K. Debattista, A. Chalmers, Advanced High Dynamic Range Imaging, first ed., AK Peters, Natick, MA, 2011.

[16] T. Capin, K. Pulli, T. Akenine-Möller, The state of the art in mobile graphics research, IEEE Comput. Grap. Appl. 28 (4) (2008) 74–84.

[17] Expert Reviews, Best battery life 2014—60 smartphones tested, http://www.expertreviews.co.uk/mobile-phones/1402071/best-battery-life-2014-60-smartphones-tested (accessed October 7, 2015).

[18] A. Rahmati, A. Qian, L. Zhong, Understanding human-battery interaction on mobile phones, in: MobileHCI'07, New York, NY, USA, 2007, pp. 265–272.

[19] D. Ferreira, A. Dey, V. Kostakos, Understanding human-smartphone concerns: a study of battery life, in: K. Lyons, J. Hightower, E.M. Huang (Eds.), International Conference on Pervasive Computing (Pervasive'11), San Francisco, CA, USA, 2011, pp. 19–33.

[20] M. Melo, HDR Video for Mobile Devices, PhD, Universidade sde Trás-os-Montes e Alto Douro, 2015.

[21] A. Rempel, W. Heidrich, H. Li, R. Mantiuk, Video viewing preferences for HDR displays under varying ambient illumination, in: Proceedings of the 6th Symposium on Applied Perception in Graphics and Visualization (APGV'09), 2009, pp. 45–52.

[22] M. Melo, M. Bessa, K. Debattista, A. Chalmers, Evaluation of tone-mapping operators for HDR video under different ambient luminance levels, Comput. Graph. Forum 34 (8) (2015) 38–49.

[23] C. Meira, M. Melo, M. Bessa, L. Barbosa, A. Chalmers, Performance of HDR video in mobile devices, in: HDRi2013—First International Conference and SME Workshop on HDR Imaging, 2013, pp. 17–21.

[24] Gohdr.com, Encoder, 2015, http://gohdr.com/website/encoder/ (accessed October 7, 2015).

[25] J. McNamee, J. Hatchett, K. Debattista, A. Chalmers, Live HDR video streaming on commodity hardware, in: Applications of Digital Image Processing XXXVIII, vol. 9599, 2015.

CHAPTER 10

Gaze-Dependent Tone Mapping for HDR Video

R. Mantiuk
West Pomeranian University of Technology, Szczecin, Poland

Contents

1. INTRODUCTION

The human visual system (HVS) is sensitive to contrast in luminance ranging from a 10^{-6}cd/m^2 (objects viewed under illumination from the stars) to 10^{8}cd/m^2 (objects viewed on a bright sunny day) [1]. However, momentary dynamic range is limited to 4 orders of magnitude. In this range of luminance, a human can see details and at the same time, experiences lower luminance as noise and higher luminance as over-saturated areas. To extend the dynamic range, the HVS has the ability to adapt to changes in the ambient luminance and move the "detailed vision" window along the luminance range of a scene. Interestingly, it mainly adapts to an area covering approximately 2–4 degrees of the viewing angle around the gaze direction [2, 3]. Other areas of the scene, observed not in foveal but in para-foveal, and peripheral regions, have significantly less impact on the adaptation level, although, a human frequently changes gaze direction (even a hundred times per second) and tries to adapt to different regions. As the

High Dynamic Range Video
http://dx.doi.org/10.1016/B978-0-12-809477-8.00010-8

process of the luminance adaptation is slower than changes of gaze direction, the HVS is permanently in the *maladaptation state*, in which the adaptation luminance is changing toward a target value, but never reaches this value, because in the meantime, the target is changed.

We dynamically reproduce the process of maladaptation using the scene data stored in high dynamic range (HDR) video. We use an eye tracker to capture the gaze direction of a human observer. Then, we compute the temporal adaptation luminance and use its value to display the HDR image on the low dynamic range (LDR) display. The global tone mapping parameterized by the temporal adaptation can be applied to HDR frames to obtain perceptually correct and plausible tone compression results. For every pixel in the HDR frame, we use the same compression curve which changes over time, following the new values of the adaptation luminance. This technique, maintaining the advantages of the global compression, like lack of video artefacts, halos, or perceptual plausibility of visualization, and simultaneously retains visibility of details in the high dynamic range regions. Advanced local tone mapping algorithms apply different compression curves for every pixel in an HDR image. These techniques preserve visibility of details in the output LDR image, although the excess of details can be distracting for an observer and interpreted as unnatural. The GDTMO technique reveals all the details in the HDR image and at the same time, retains noisy and over-saturated areas at the peripheral region of vision.

We call this approach a gaze-dependent tone mapping operator (GDTMO). This concept was first proposed in Mantiuk and Markowski [4]. In this work, we extend the application from static images to the HDR video. We also introduce a new model of local adaptation proposed in Vangorp et al. [2] and modify the compression curve.

In Section 2 we present an overview of the eye tracking technologies and their limitations related to the computer imaging applications. In Section 3 we explain details of GDTMO and introduce its real time implementation. Section 4 presents results of the tests of this approach and discusses its limitations. We conclude the chapter and suggest directions for further work in the last section.

2. BACKGROUND

Advanced *tone mapping techniques* were developed to minimize the deterioration of brightness, contrast, and/or perceptual appearance of the HDR images displayed on the LDR displays (see [1] for comprehensive reviews). However, most of the TMOs assume that an observer adapts to luminance of the individual pixels [5], rather than take into account the temporary adaptation to luminance caused by changes of the view direction [6]. This "nonperceptual" approach together with the local modification of the tone curve causes the unnatural appearance of the HDR content after the tone compression. We argue that more plausible results can be achieved using the eye tracking.

The devices called *video eye trackers* [7, 8] employ the pupil-corneal reflection (P-CR) technique [7]. An eye tracker usually consists of an IR camera and an IR light source, which are directed at the eye. The camera captures the image of the eye with the dark circle of the pupil and the bright corneal glint, which is a reflection of the infrared light from the outer surface of the cornea. The pupil follows the gaze direction during eye movement, while the corneal reflection stays in the same position. The relative position between the reflection and the center of the pupil is used to estimate the gaze direction.

Eye trackers capture two types of eye movements, *saccades* and *smooth pursuits* [9]. A smooth pursuit is active when eyes track a moving target and are capable of matching its velocity. A saccade represents a rapid eye movement used to reposition the fovea to a new location, which lasts from 10 to 100 ms. However, the main goal of the eye tracking is to estimate a single location an observer intends to look at for a longer time. This process is known as a *visual fixation*. A point of fixation can be estimated, for example, as a location where saccades remain stable for a 200–400 ms on the most significant areas of an image (called region-of-interest, ROI) [10].

The *accuracy* of eye trackers is defined as the average distance between the real stimuli position and the measured gaze position [9]. A typical video eye tracker is capable of estimating the gaze point on a display screen with accuracy close to 1 degree of the viewing angle (a circular region of roughly 80 pixels in diameter observed on a 22-in. display of 1960 × 1050 pixel resolution from a 60-cm distance). These are rather mediocre results, which makes the estimation of the fixation point challenging. Moreover, the accuracy is prone to higher temporal errors, even when using the state-of-the-art fixation techniques like velocity-threshold identification (I-VT) or dispersion threshold identification (I-DT). More accurate solutions exist, but they require a database of the potential fixation target [11], which is particularly difficult to built for video materials [12].

All these observations suggest that the eye tracker cannot be assumed as a device, which captures exact gaze direction. Rather, it estimates an area of the visual attention. The gaze-dependent tone mapping technique must comply with the eye tracking inaccuracies.

3. GAZE-DEPENDENT TONE COMPRESSION

The complete gaze-dependent tone compression system is illustrated in Fig. 1. The observer's gaze direction is captured by tracker. Then, the luminance of the pixels surrounding the gaze point (taken from a video frame) is used to estimate the adaptation luminance (L_a) (see Section 3.1). The current L_a value is used to build a global tone compression curve and compress the HDR content (see Section 3.3). Finally, the output R, G, and B color values are computed, and desaturated using the exponential rule from Mantiuk et al. [13] and the LDR frame is displayed on the screen. This procedure is repeated for consecutive frames in the video. Additionally, L_a is smoothly modified from the previous to the next frame, using the temporary adaptation rules (see Section 3.2).

3.1 Spacial Extent of Visual Adaptation

To compute the spacial extent of the visual adaptation, we use a model proposed in Vangorp et al. [2]. This model estimates the adaptation

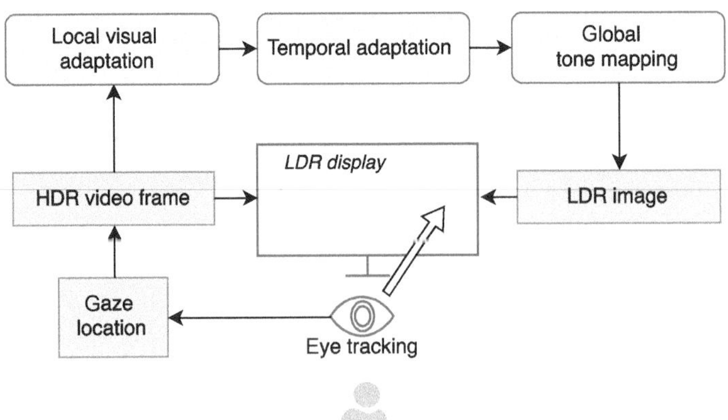

Fig. 1 The diagram of the gaze-dependent tone mapping (GDTMO) pipeline.

luminance, applying the OTF filter in the linear luminance domain and two Gaussian pooling filters in nonlinear domains:

$$L_a = \alpha n_1^{-1}(n_1(L_O) * g_{\sigma_1}) + (1-\alpha)n_2^{-1}(n_2(L_O) * g_{\sigma_2}), \quad (1)$$

where L_a denotes the output adaptation luminance, $\alpha = 0.654$, $*$ is the convolution operator, parameters of the Gaussian kernels g are $\sigma_1 = 0.428°$ and $\sigma_2 = 0.0824°$. n_1 and n_2 are nonlinearities that can be approximated with sigmoidal function:

$$n(L_O) = a\frac{L_O^c}{b + L_O^d}. \quad (2)$$

The parameters a, b, c, d were estimated by training the model on empirical data obtained from the perceptual experiments. The authors propose using 3.46, 70.2, 2.38, 2.37 for n_1, and 2.19, 66.8, 2.37, 2.37 for n_2, respectively. L_O is the retinal luminance given by the convolution of a HDR frame luminance with the OTF proposed by Deeley et al. [14] (with assumed pupil diameter 4 mm).

Fig. 2 shows the response of the model to stimuli containing a 0.2° disk displayed at luminance ranging from 1 to 10,000 cd/m^2. An example of a local adaptation map computed for the natural image is presented in Fig. 3.

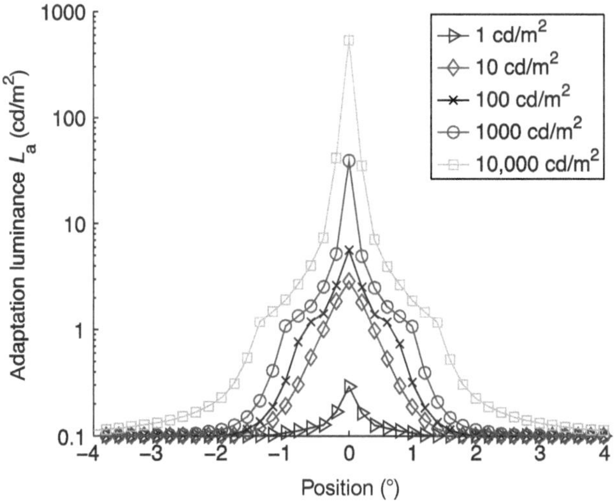

Fig. 2 A response of the local adaptation model.

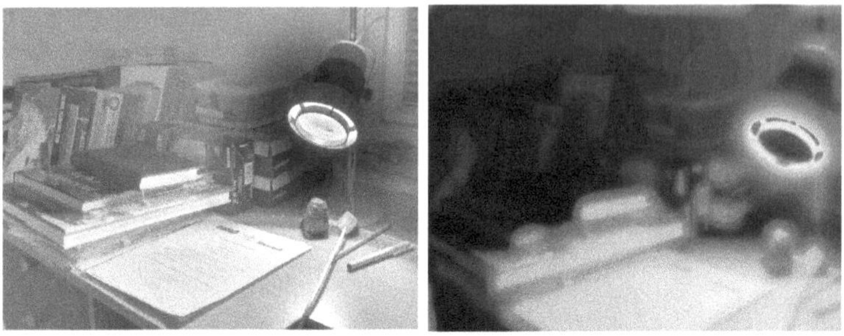

Fig. 3 Tone mapped HDR image (*left*) and corresponding false color map with the local adaptation values (*right*).

3.2 Temporal Adaptation to Light and Dark

The luminance conditions can change drastically over time. The HVS reacts to such changes through the temporal adaptation process, which differs depending on whether we adapt to light or to darkness, and whether we perceive mainly using rods (during nighttime) or cones (during daytime). In the tone mapping algorithm, the adaptation can be modeled by varying the adaptation luminance value over time. Ferwerda et al. [15] presented a computational model of changes in luminance sensitivity over time using Ward's scaling tone mapping approach [16]. Pattanaik et al. [17] proposed a complex photoreceptor-based, time-dependent, retinal adaptation mechanisms for both cones and rods.

In GDTMO we implemented a simplified model, which is based on the mentioned approaches, proposed in Krawczyk et al. [18]:

$$\hat{L}_a^{\text{new}} = \hat{L}_a + (\hat{L}_{\text{HDR}} - \hat{L}_a)(1 - e^{-\frac{T}{\tau}}), \tag{3}$$

where L_a and L_a^{new} are the previous (from previous video frame) and new adaptation luminance, respectively, L_{HDR} denotes the luminance of the HDR frame at location of the gaze point, and $\hat{L} = log_{10}(L)$. T is the time step between the display of two frames, and the τ constant defines the speed of the adaptation process. This speed is different for cones and rods; to account for this, we use the following interpolation:

$$\tau(L_a) = \sigma(L_a) * \tau_{\text{rod}} + (1 - \sigma(L_a)) * \tau_{\text{cone}}, \tag{4}$$

where $\tau_{\text{rod}} = 0.4\,\text{s}$, $\tau_{\text{cone}} = 0.1\,\text{s}$, and σ models the sensitivity of rods with the following equation:

$$\sigma(L_a) = \frac{0.04}{0.04 + L_a}. \tag{5}$$

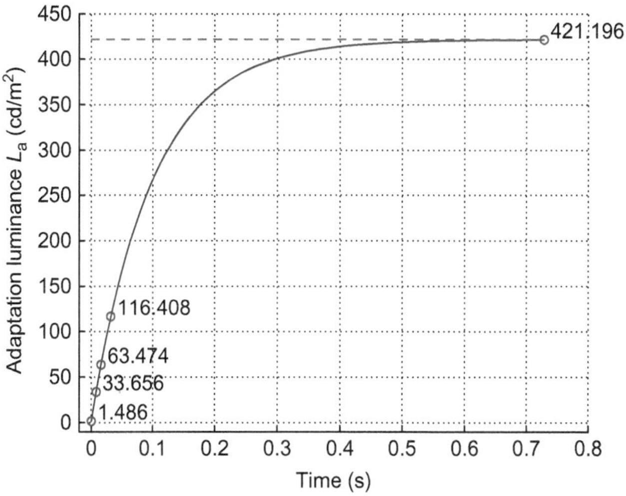

Fig. 4 Example transition of temporary adaptation value over time. The circles correspond to 1st, 2nd, 3rd, 5th, and 93th frame of video displayed on a 120-Hz display (each frame shown every 8 ms) (see Fig. 6 for the screenshots).

The presented model describes adaptation to light, in which the observer adapted to a dark area turns his eyes to the bright areas of the HDR frame. Fig. 4 shows an example response of the adaptation luminance over time.

We simulate the reverse process—adaptation to the dark as the linear decrease of adaptation luminance. It is slow process, which does not require any nonlinear adaptation model, because this change in practice would be imperceptible to an observer [18]. Furthermore, we artificially accelerate the adaptation by two orders of magnitude. In the real world, the adaptation to dark takes up to tens of minutes and preserving perceptual correctness would be impractical in GDTMO. After quick adaptation to light (about a second), one would wait for more than a minute to see any changes caused by the adaptation to dark.

3.3 Tone Compression

For HDR tone compression we apply the modified Naka-Rushton equation [19], which models the intensity-response function of the distal retinal neurones in HVS:

$$L_{\mathrm{LDR}} = \frac{L_{\mathrm{HDR}}^n}{L_{\mathrm{HDR}}^n + \sigma(L_a)^n}, \tag{6}$$

where L_{LDR} denotes the output low dynamic range luminance, L_{HDR} is a luminance of HDR pixels, and n is a constant between 0.7 and 1 (we

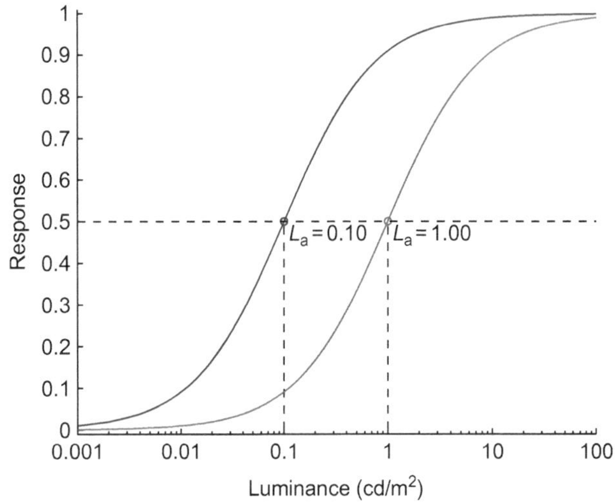

Fig. 5 Naka-Rushton photoreceptor response model for two example adaptation levels.

assumed $n = 1$ in our implementation). The semisaturation constant $\sigma(L_a)$ controls the translation of the response curve along the log-luminance axis, as shown in Fig. 5. Following Vangorp et al. [2], we assumed that this factor is equal to the adaptation luminance: $\sigma(L_a) = L_a$.

The mechanism of visual adaptation is responsible for constant contrast perception in varying luminance conditions and in a direct way can be adapted for compression of HDR values. Additionally, the Naka-Rushton equation correctly mimics the perceived brightness in a maladaptation state, because it was modeled in an experiment with pulsing stimuli [20]. It is consistent with natural HVS behavior, as we use saccadic movements to scan the whole scene, thus, the stimuli fields are not steady, but pulsed.

4. TESTBED AND RESULTS

GDTMO was implemented in Matlab using a set of GPU shaders controlled by the Psychtoolbox[1] library. The shaders are responsible for tone compression, color desaturation, and final image display. The software was executed on a 2.8-GHz Intel i7 930 CPU equipped with an NVIDIA GeForce 580 GTI 3072MB graphics card. To capture gaze direction, we use 250 Hz P-CR RED250 eye tracker controlled by the proprietary SMI

[1] http://www.psychtoolbox.org/HomePage.

iViewX software (version 2.5) running on a dedicated PC. RED250 eye tracker was mounted under a 22-in. Dell E2210 LCD display with the screen dimensions 47.5 × 30 cm, and the native resolution 1680 × 1050 pixels (60 Hz).

Fig. 6 presents a set of output images generated by GDTMO. An observer changes his/her gaze inducing modification of the target adaptation luminance from 0.43 to 421.2 cd/m^2. The temporary adaptation smoothly changes between these values. The presented images correspond to levels of the adaptation luminance depicted as magenta circles in Fig. 4.

Fig. 6 Images viewed by an observer adapted to two different areas of the scene indicated by the red crosses (dark gray in print versions). In the top-left image the adaptation luminance is close to 0.43 cd/m^2, in the remaining ones it changes over time to reach the target 421.2 cd/m^2.

5. CONCLUSIONS AND FUTURE WORK

We have shown how information about gaze direction can be used to tone map HDR video in real time. The proposed model is based on local and temporary adaptation to luminance and perceptual compression of the luminance range. The eye tracker is used to capture observers' gaze location and compute the spatial extend of visual adaptation. The local adaptation value modified over time to account for the maladaptation state is applied by the tone mapping technique to adjust the global compression curve to the current brightness.

In future work, we plan to test the perceptual plausibility of this solution. It would be interesting to evaluate if humans would not experience changes in the displayed video as unnatural, or in other words, if they would accept the time and content-dependent changes of the global luminance level. To perform such an experiment, a fast display offering the wide viewing angle will be the most suitable. We plan to conduct the experiment based on the head-mounted displays, as well as use displays that show more than 120 frames per second.

ACKNOWLEDGMENTS

This work was partially funded by the Polish National Science Centre (decision number DEC-2013/09/B/ST6/02270).

REFERENCES

[1] E. Reinhard, W. Heidrich, P. Debevec, S. Pattanaik, G. Ward, K. Myszkowski, High Dynamic Range Imaging, Acquisition, Display, and Image-Based Lighting, second ed., Morgan Kaufmann, San Francisco, CA, 2010.

[2] P. Vangorp, K. Myszkowski, E.W. Graf, R.K. Mantiuk, A model of local adaptation, ACM Trans. Graph. 34 (6) To appear (Proceedings of ACM SIGGRAPH Asia 2015).

[3] D.C. Hood, Lower-level visual processing and models of light adaptation, Annu. Rev. Psychol. 49 (1998) 503–535.

[4] R. Mantiuk, M. Markowski, Gaze-Dependent Tone Mapping, vol. Lecture Notes in Computer Science (Proc. of ICIAR'13 Conference) 7950, 2013, pp. 426–433.

[5] R. Mantiuk, K. Myszkowski, H.-P. Seidel, A perceptual framework for contrast processing of high dynamic range images, ACM Trans. Appl. Percept. 3 (3) (2006) 286–308.

[6] R.K. Mantiuk, G. Krawczyk, R. Mantiuk, H.P. Seidel, High dynamic range imaging pipeline: perception-motivated representation of visual content, in: Proc. of SPIE 6492 Human Vision and Electronic Imaging XII, 2007, p. 649212.

[7] T.A. Duchowski, Eye Tracking Methodology: Theory and Practice, second ed., Springer, New York, 2007.

[8] A. Nowosielski B. Bazyluk R. Mantiuk, M. Kowalik, Do-it-yourself eye tracker: low-cost pupil-based eye tracker for computer graphics applications, in: Proc. of MMM 2012, Lecture Notes in Computer Science, vol. 7131, 2012, pp. 115–125.

[9] K. Holmqvist, M. Nystrom, R. Andersson, R. Dewhurst, H. Jarodzka, J. van de Weijer, Eye Tracking: A Comprehensive Guide to Methods and Measures, first ed., Oxford University Press, USA, 2011.

[10] D.D. Salvucci, J.H. Goldberg, Identifying fixations and saccades in eye-tracking protocols, in: Proceedings of the 2000 Symposium on Eye Tracking Research and Applications (ETRA), New York, 2000, pp. 71–78.

[11] R. Mantiuk, B. Bazyluk, R.K. Mantiuk, Gaze-driven object tracking for real time rendering, Comput. Graph. Forum 32 (2) (2013) 163–173.

[12] B. Bazyluk, R. Mantiuk, Gaze-driven object tracking based on optical flow estimation, in: Computer Vision and Graphics, Springer, Berlin, 2014, pp. 84–91.

[13] R. Mantiuk, R. Mantiuk, A. Tomaszewska, W. Heidrich, Color correction for tone mapping, Comput. Graph. Forum 28 (2) (2009) 193–302 (Proc. of EUROGRAPH-ICS 2009).

[14] R.J. Deeley, N. Drasdo, W.N. Charman, A simple parametric model of the human ocular modulation transfer function, Ophthal. Physiol. Optics 11 (1) (1991) 91–93.

[15] S.N.J.A. Ferwerda, Pattanaik, P. Shirley, D.P. Greenberg, A model of visual adaptation for realistic image synthesis, in: Proc. of SIGGRAPH'96, 1996, p. 249258.

[16] G. Ward, A contrast-based scalefactor for luminance display, Graphics Gems IV, 1994.

[17] S.N. Pattanaik, J. Tumblin, H. Yee, D.P. Greenberg, Time-dependent visual adaptation for fast realistic image display, in: Proc. of SIGGRAPH'00 2000, pp. 47–54.

[18] G. Krawczyk, K. Myszkowski, H. Seidel, Perceptual effects in real-time tone mapping, in: Proceedings of the 21st Spring Conference on Computer graphics, Budmerice, 2005, pp. 195–202.

[19] K.-I. Naka, W.A. Rushton, S-potentials from luminosity units in the retina of fish (cyprinidae), J. Physiol. 185 (3) (1966) 587–599.

[20] S. Ferradans, M. Bertalmio, E. Provenzi, V. Caselles, An analysis of visual adaptation and contrast perception for tone mapping, IEEE Trans. Pattern Anal. Mach. Intell. 33 (10) (2011) 2002–2012.

PART 4

Applications and Uptake

CHAPTER 11

Improving Tracking Accuracy Using Illumination Neutralization and High Dynamic Range Imaging

N. Ladas*, Y. Chrysanthou*, C. Loscos†
*University of Cyprus, Nicosia, Cyprus
†University of Reims Champagne-Ardennes, Reims, France

Contents

1. INTRODUCTION

Object recognition and tracking is a class of video processing algorithms that identify targets as they appear in a video (recognition) and then follow their location for as long as they remain in view (tracking).

When the sensor's dynamic range cannot accommodate the full dynamic range of the scene, parts of the image appear completely dark (undersaturated) or washed out (oversaturated) and important information may be lost. In the case of object detection, information in the saturated regions could be useful in the identification process, as it might contain features that describe the object that we aim to identify. In tracking, saturated regions can cause the tracking algorithm to fail when the target enters these regions.

High Dynamic Range Video
http://dx.doi.org/10.1016/B978-0-12-809477-8.00011-X

Robust object recognition and tracking is a challenging task even in well exposed, constant-illumination scenes. Rapid changes in illumination conditions, such as lights turning on/off or clouds obscuring the sun can cause significant changes in the appearance of the captured scene that, in turn, cause significant reduction in tracking accuracy. In addition, illumination effects, such as shadows and specular highlights, can perturb algorithm efficiency.

High dynamic range (HDR) imaging can alleviate these problems by extending the dynamic range that is captured. Simply employing HDR inputs in unmodified conventional techniques can be beneficial, as less information is lost due to saturation, while further performance gains are possible when techniques are designed to take advantage of HDR information.

This chapter presents an in-depth discussion of the technique by Ladas et al. [1, pp. 51–54] that improves the accuracy of object tracking algorithms by employing a custom HDR capture methodology, and an image neutralization algorithm designed to take advantage of HDR information. While the work presented is used as a guiding example, we expand the relevant related work and give insights to the implementation specifics and directions for future work.

2. OVERVIEW

The work presented in this chapter aims to improve the accuracy of object tracking algorithms by employing HDR imaging and a method for scene illumination normalization. The algorithm takes as input an HDR video stream where the illumination can have potentially significant variation in intensity. Additionally, a set of HDR fisheye images is also used that capture the scene illumination (incoming light). By analyzing the scene illumination the method is able to reverse the illumination effects that appear on the input images and produce an illumination-neutral image stream. Tracking objects using the illumination neutral images is shown to be much more robust.

As the HDR video stream of the tracked area is captured, a separate low dynamic range (LDR) camera, mounted with a fisheye lens, takes snapshots of the scene's incoming illumination and assembles them in the form of an HDR environment map. This is done at regular time intervals (typically every few seconds). Given an estimate of the scene geometry visible through each pixel, the incoming illumination from the current HDR environment map is used to compute in real-time the scene lighting for each point. Using this, lighting disparities, such as shading and shadows are removed, producing an illumination-neutral video stream that can be used as input by any tracking algorithm.

3. ACQUISITION OF THE SCENE ILLUMINATION

3.1 Fisheye Images

To acquire the scene illumination, a standard SLR camera is used, mounted with a fisheye lens pointing upward. At this configuration, one shot captures a hemispherical view of the scene (typically up to 360 × 90 degrees) that includes most of the area contributing to the incoming illumination in typical scenes (see Fig. 1). The standard assumption is made that the main sources of light in the environment are relatively far from the observed part of the scene. If this assumption does not hold, fisheye images must be

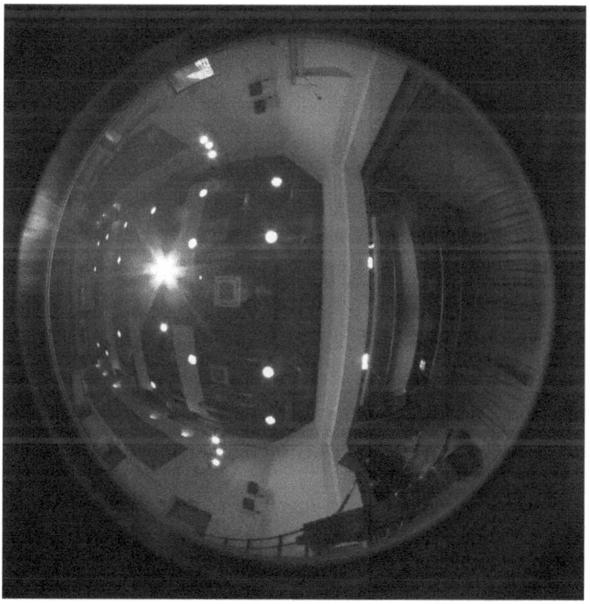

Fig. 1 Image captured with a camera equipped with a fisheye lens. The camera points upward to capture the illumination arriving from the light sources of the scene.

captured close to the point of interest in the scene, and sampled in multiple locations to form a light field [2, pp. 31–42].

Directly photographing light sources are usually not possible, as the regions of the image depicting the light sources will be saturated. Standard LDR to HDR assembly [3, pp. 369–378] was used to properly capture the fisheye images. For very bright light sources, such as the sun, it may still not be possible to capture the image directly with conventional cameras. In such cases, specialized capture techniques must be employed.

One possible approach is to use a light probe to capture the incoming illumination at a specific location. A light probe is a highly reflective metallic sphere that reflects the illumination arriving from nearly every direction. Because we are photographing the probe instead of the light sources directly, probe reflectance must be determined prior to using a calibration procedure [4, pp. 189–198]. Because the probe reflects less light than it receives, this indirection is often enough to avoid saturation of the image sensor. In very strongly lit scenes, such as when taking a light probe of the sky in a sunny day, the light reflected by the probe will still saturate the camera sensor. An approach for dealing with this problem was described by Debevec and Lundgren [5, pp. 1–11] which involved photographing a separate diffuse (mat) gray ball. The authors calibrated the reflectance of the gray ball against a known illuminant and used it to capture the illumination of the sun. A conventional light probe was then used to capture the illumination from the rest of the sky.

Another approach is to use neutral density (ND) filters to block a portion of the incoming illumination. ND filters are darkened plastic membranes that attach to the camera lens, typically quantified by the amount of illumination they block (expressed in f-stops). For accurate results, the ND filter must be calibrated, as it may not block the exact amount of illumination advertised or it may behave evenly across its surface. The work by Stumpfel [6, pp. 145–150] describes an approach that performs this calibration.

For highly accurate scene illumination capture, trichromatic (e.g., red-green-blue) light sampling may not be sufficient and the whole visible spectrum must be captured [7, pp. 1–12]. This, however, requires specific, high-cost equipment and cumbersome capture procedures. For all cases considered in this chapter, standard LDR to HDR acquisition was sufficient.

3.2 Environment Map Representation

The captured fisheye images must be converted to a known representation. A commonly used representation is equirectangular (latitude-longitude), as shown in Fig. 2.

Equirectangular projection maps the surface of a sphere onto a plane. The horizontal direction represents latitude and the vertical direction longitude. This representation is widely used as it allows for easy, intuitive access of the data. This is because pixels locations (x, y) directly translate into spherical coordinates (θ, ϕ). To successfully project the captured fisheye image to the altitude-longitude projection, the fisheye lens must be calibrated with a technique such as Scaramuzza et al. [8, pp. 5695–5701]. The fisheye lens captures up to half a hemisphere and as a result, when projected to an equirectangular image, half of the image is black. Equirectangular images representing scene illumination are often referred to as environment maps (a term that we will use for the remainder of the chapter).

To capture changing light conditions, such as lights turning on and off, the scene illumination must be captured regularly. In the current experiment, a standard SLR camera, controlled remotely by a laptop running freely available software (http://gphoto.sourceforge.net/), was used to capture the fisheye images at regular time intervals. Due to the bracketing, the transfer from camera to computer and the assembly of the images, it was not possible to produce more than one fisheye HDR image every few seconds. For environments where the illumination conditions change more frequently, it would be desirable to use an HDR video camera.

Fig. 2 A fisheye image converted to the latitude-longitude representation. The fisheye image only captures the appearance of the top hemisphere and as a result, the bottom half of the resulting latitude-longitude image is black.

4. PRECOMPUTED RADIANCE TRANSFER

Typically, to correctly calculate the illumination arriving at each scene location, we must sample from there the environment map, using a suitable distribution and taking into account obstructions and interreflections. This is a time-consuming task that is not well suited to the real-time nature of object detection/tracking. This is a common problem in real-time graphics and it is commonly solved by making simplifying assumptions that trade accuracy for efficiency, such as ignoring interreflections and precomputing visibility for static scene components.

A good compromise between accuracy and efficiency is the precomputed radiance transfer (PRT) technique [9, pp. 527–536, 10, pp. 131–160]. In PRT, reflectance and visibility at key scene locations (typically geometry vertices) are calculated using Monte Carlo integration and projected into spherical harmonic (SH) coefficients. The incoming illumination is assumed to be coming from far enough, thus having the same relative origin and direction when seen from any point in the scene. The illumination is also encoded as one set of SH coefficients. Encoding the visibility is a slow process and must be done offline (once per scene) while encoding the incoming illumination is done in real-time. In our case we assumed mostly diffuse objects, therefore the reflectance is a cosine term that can be encoded into SH during preprocessing. At runtime, the illumination arriving at scene point x can be calculated as the sum of the products of the SH coefficients of incoming illumination L and the transfer function (visibility + cosine term) T:

$$I(x) = \sum_{i=0}^{N} L_i T_i(x)$$

where n is the number of SH coefficients used. Increasing the number of coefficients leads to better results at the cost of using more memory. Typically, 16 SH coefficients provide good lighting results in low-frequency illumination scenarios. The light reflected toward the camera is then calculated as the product of incoming illumination and scene reflectance:

$$R(x) = p(x)\, I(x)$$

Standard PRT implementations calculate the visibility for each geometry vertex. This approach has two limitations. If the scene is overly complex, the processing time and memory requirements for the visibility computation increase dramatically. On the other hand, if the scene consists of crude geometry with few vertices, visibility will be undersampled and the lighting quality will suffer. In many tracking scenarios, the camera is static (e.g., CCTV cameras). We exploit this fact, and compute the scene visibility for each screen pixel instead of each scene vertex. We perform an initial ray-cast step that traces rays from the camera toward the scene. At each intersection, we trace random rays from the intersection point toward the rest of the scene in order to estimate the visibility. This approach works well in complex scenes, because the number of visibility computations is always constant (equal to the number of pixels). In simple scenes, this approach naturally multisamples the scene and produces better lighting results than the per-vertex computation. In the cases where the camera is not static, the system can revert to per-vertex computation.

A limitation of PRT is that it requires the scene's geometric information to be known, which may not be the case for real world scenes. When the scene geometry is not available, we work around this limitation by calculating the incoming illumination on a hypothetical flat surface. Because many tracking scenarios consist of tracking objects on flat surfaces (e.g., tracking pedestrians in a plaza or the cars on a road), this assumption works quite well. When, however, the goal is to maximize the accuracy of the algorithm, the scene geometry must be provided. Digitizing a real scene's static geometric information is a well-studied field and many approaches are available [11, pp. 834–849].

5. NEUTRALIZING ILLUMINATION

5.1 Sampling Environment Maps

Sampling and encoding the visibility is a slow process and is done once per scene and offline. Sampling the illumination given by the environment map can be performed in real time, as long as the number of samples used is small. Efficient sampling can be achieved using relatively few samples by employing importance sampling techniques [12, p. 33]. The key concept on which importance sampling techniques are based is to allocate more samples toward regions in the image that have more brightness, as these regions will likely contribute more to the scene illumination.

5.2 Neutralizing

Sampling the illumination in real time is beneficial as it allows for more than one environment map to be used during the tracking operation. This is important in cases where the scene illumination changes abruptly during tracking. Because our system captures both the video stream and a sequence of environment maps concurrently, it is able to better react to illumination changes. Acquiring and sampling the environment maps can take a few seconds, and so multiple video frames are associated with the same environment map. Each new environment map is sampled and encoded into SHs and the scene illumination is recalculated. After the illumination is calculated, we subtract it from the input video. Because scene materials are assumed to be diffuse, we can estimate the reflectance ρ of the scene's objects by dividing each pixel in the video frame by the illumination calculated for the scene location associated with that pixel:

$$p(x) = pixel(i) / I(x)$$

After the reflectance is calculated, we obtain the illumination-neutral video by lighting the scene objects with ambient lighting. In order to easily associate video pixels with scene locations, we align the virtual camera used in the PRT system with the real camera.

6. IMPROVING TRACKING ACCURACY

To evaluate how our neutralization scheme affects tracking accuracy, we compared the tracking results of tracking-learning-detection (TLD) [13, pp. 49–56], a high-end real time tracking algorithm, using standard video versus illumination neutralized video as input. TLD is resilient against changes in global illumination and can adapt, given enough training input, to strong changes in local illumination. In TLD the user is tasked with specifying the initial position of the target object and no object detection is performed. Alternatively, the initial position of the target could be determined using an object detection method such as [14, pp. 1627–1645].

The test scenario consisted of tracking a small object on a flat surface. As the object moved, we altered the illumination conditions by switching lights on/off and obscuring the light coming from the windows. The goal was to capture different illumination conditions (well-lit, under-lit, directional lighting) and also to have rapid changes in illumination. At each illumination change, we captured a new environment map. Fig. 3 shows in columns 1 and 2, the results when using an unprocessed video, and in columns 3 and 4, the results when using the illumination neutral video produced by our algorithm.

Successful tracking is shown in each image as a circle over the target. When sharp illumination changes occur, tracking fails for the unprocessed

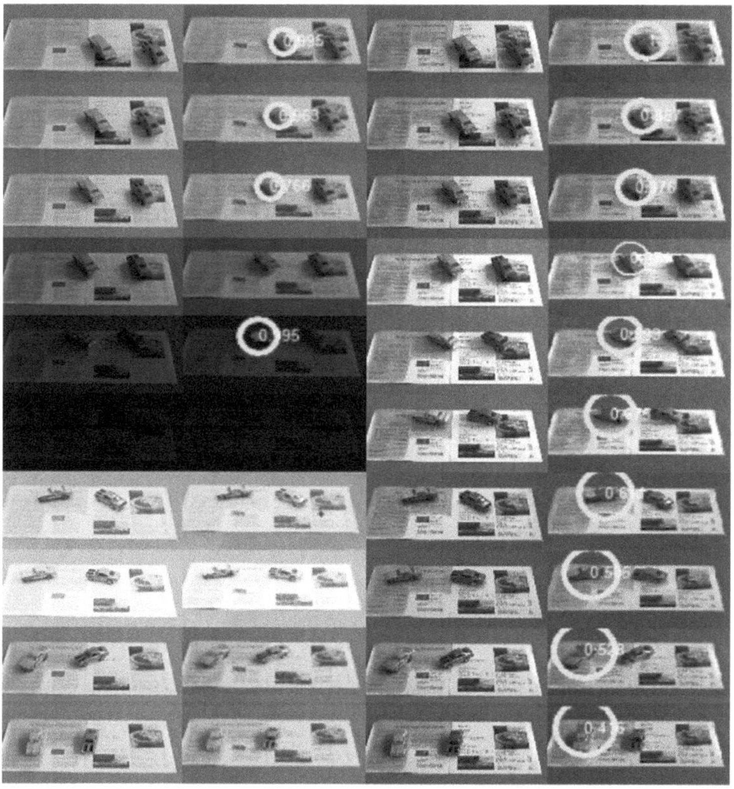

Fig. 3 The *first column* shows the original input stream while the *second column* shows the tracker's output for this stream. The *third column* shows the illumination neutral stream produced by the algorithm. The *fourth column* shows the tracking result for the illumination neutral input. Tracking accuracy is substantially improved for illumination neutral inputs.

video (frames 4, 6–10). The illumination-neutral video maintains a constant scene appearance by removing illumination effects and as result tracking continues unhindered.

Because the TLD tracker operates on LDR input, we had to convert the HDR input frames into LDR images. To do this, we selected a fixed exposure setting from the LDR stack used to create the HDR frames. Another approach would be to tone map the HDR and use the result as input for the tracker. We tested our data using the tone-mapping procedure described in Reinhard and Devlin [15, pp. 13–24] and gave similar results to using a fixed exposure. Evaluating the effect of tone mapping on tracking and other computer vision methods is an open direction for further work.

7. CONCLUSIONS

This chapter presented a method for enhancing the accuracy of a high-end object tracker. The method used HDR inputs instead of conventional video, and additionally captured the scene illumination in the form of environment maps. Using this information, the appearance of the inputs was normalized and the effects of sudden light changes, that dramatically darkened or brightened the scene, were removed. This translated to concrete improvements in tracking performance.

The algorithm presented here is relatively simple, but still manages to improve tracking performance significantly. We believe that the results presented here are indicative of the possible performance gains of using HDR-enabled object trackers, although more thorough testing is required to solidify this claim. At the same time, there is very little literature involving HDR imaging and tracking, object recognition, or similar disciplines, which leads us to believe that the area is good target for future work.

REFERENCES

[1] N. Ladas, Y. Chrysanthou, C. Loscos, Improving tracking accuracy using illumination neutralization and high dynamic range imaging, in: HDRi2013—First International Conference and SME Workshop on HDR Imaging, vol. 7, 2013, pp. 2–6.

[2] M. Levoy, P. Hanrahan, Light field rendering, in: Inproceedings of the 23rd Annual Conference on Computer Graphics and Interactive Techniques—SIGGRAPH '96, ACM Press, New York, NY, 1996, pp. 31–42, http://portal.acm.org/citation.cfm?doid=237170.237199 (accessed December 12, 2013).

[3] P.E. Debevec, J. Malik, Recovering high dynamic range radiance maps from photographs, in: Inproceedings of the 24th Annual Conference on Computer Graphics and Interactive Techniques. SIGGRAPH '97, ACM Press/Addison-Wesley Publishing Co., New York, NY, 1997, pp. 369–378.

[4] P. Debevec, Rendering synthetic objects into real scenes: bridging traditional and image-based graphics with global illumination and high dynamic range photography, in: Inproceedings of the 25th Annual Conference on Computer Graphics and Interactive Techniques. SIGGRAPH '98, ACM, New York, NY, 1998, pp. 189–198.

[5] P. Debevec, T. Lundgren, Estimating surface reflectance properties of a complex scene under captured natural illumination, ACM Trans. Graph. 19 (2004) 1–11.

[6] J. Stumpfel, Direct HDR capture of the sun and sky, in: Inproceedings of the 3rd International Conference on Computer Graphics, Virtual Reality, Visualisation and Interaction in Africa—AFRIGRAPH '04, ACM Press, 2004, p. 145, http://portal.acm.org/citation.cfm?doid=1029949.1029977 (accessed September 12, 2014).

[7] J.T. Kider, A framework for the experimental comparison of solar and skydome illumination, ACM Trans. Graph. 33 (6) (2014) 1–12.

[8] D. Scaramuzza, A. Martinelli, R. Siegwart, A toolbox for easily calibrating omnidirectional cameras, in: 2006 IEEE/RSJ International Conference on Intelligent Robots and Systems, 2006, pp. 5695–5701.

[9] P.-P. Sloan, J. Kautz, J. Snyder, Precomputed radiance transfer for real-time rendering in dynamic, low-frequency lighting environments, ACM Trans. Graph. 21 (3) (2002) 527–536.
[10] M.P.B. Slomp, M.M. Oliveira, D.I. Patricio, A gentle introduction to precomputed radiance transfer, Revista De Informática Teórica E Aplicada 13 (2) (2006) 131–160.
[11] J. Engel, J. Stückler, D. Cremers, LSD-SLAM: large-scale direct monocular SLAM, in: Lecture Notes in Computer Science (including subseries Lecture Notes in Artificial Intelligence and Lecture Notes in Bioinformatics), 8690 LNCS (Part 2), 2014, pp. 834–849.
[12] P. Debevec, A median cut algorithm for light probe sampling, in: ACM SIGGRAPH 2008 Classes, ACM, New York, NY, 2008, p. 7154, http://dl.acm.org/citation.cfm?id=1401176 (accessed November 19, 2014).
[13] Z. Kalal, J. Matas, K. Mikolajczyk, P-N learning: bootstrapping binary classifiers by structural constraints, in: 2010 IEEE Computer Society Conference on Computer Vision and Pattern Recognition, 2010, pp. 49–56.
[14] P.F. Felzenszwalb, Object detection with discriminative trained part based models, IEEE Trans. Pattern Anal. Mach. Intell. 32 (9) (2010) 1627–1645.
[15] E. Reinhard, K. Devlin, Dynamic range reduction inspired by photoreceptor physiology, IEEE Trans. Visual. Comput. Graph. 11 (1) (2005) 13–24.

CHAPTER 12

Factors Influencing the Widespread Uptake of HDR Video

C. Moir, A. Chalmers
University of Warwick, Coventry, United Kingdom

Contents

1. INTRODUCTION

Adoption of high dynamic range (HDR) video is defined as expenditure by firms, households, and governments on products from capture to display that embody a new technology, or bundle of new technologies that make up an HDR pipeline. Thousands of visitors to the 2016 Consumer Electronics

High Dynamic Range Video
http://dx.doi.org/10.1016/B978-0-12-809477-8.00012-1

Show witnessed the introduction of the first commercially available HDR televisions. Promoted as offering more vibrant colors, much brighter images, greater contrast, deeper blacks, and the major enhancement of 4K resolution, Sony, LG, Panasonic, and Samsung all proclaimed HDR as the technology of the future. Branded as ultra high definition (UHD), the televisions were big. The largest Sony HDR model, for example, has an 85-in. display; the smallest is 55 in. Comfortable viewing would, according to ITU-R recommendations [1], require a room that allowed over 11 ft between the viewer and the 85-in. screen. Commercial adoption of HDR video is in terms of 10 or 12 bit images. The televisions are being sold online, in specialist shops, and department stores in the United Kingdom and the United States among other countries. In televisions at least HDR video had unambiguously been adopted by some leading manufacturers.

Despite the fact that since CES16 HDR has been made widely available by leading television manufactures, content is limited for now. At the same show, Netflix and Amazon offered 4K HDR content. As of Apr. 2016, Netflix had offered viewers Season 1 of the Marco Polo series in 4K HDR. This required the viewer to join the Netflix 4K content streaming plan, priced at $12 per month, and to own an HDR TV, as well as an internet connection of at least 20 Mbps, enough for supporting 4K UHD content streams [2]. Amazon Prime, on the other hand, offered 27 4K HDR films. Again, the viewers needed to be customers of Amazon Prime subscription service and own an HDR 4K TV [3]. Warner Brothers also plans to release 35 4K HDR Blu-Ray films in 2016. All hope others will follow their lead. Video capture in HDR is less of an issue. The ARRI Alexa, Red Epic, Sony F65, and a host of other cameras are capable of capturing raw footage with a dynamic range in excess of the current televisions' capability to show it [4]. For now (summer 2016), commercially HDR video enabled displays are confined to televisions alone. HDR video is not yet commercially available on phones, tablets, or laptop computers.

Current HDR video commercial adoption is less than technologically optimal. At each stage of the HDR video pipeline, the technology in commercially available HDR televisions is inferior, in performance, to that in the best available technologies. Cameras capturing 10-bit content typically record around 10 f-stops where, with the right sensors, they could capture up to 18 f-stops, or 16 without significant noise. Much of the content is tone mapped on the camera, so a significant proportion of information is lost. Recent perceptual lossey encoding and decoding file compression techniques mean significant information and detail for human consumption does not have to be lost. The HDR branded televisions have a peak brightness of 1000 nits. SIM2 will sell you a 6000 nit monitor. Avoidance of using the best available technology in consumer electronics is the exception, rather than the rule. A more typical story over the last three decades is one technology embodied in one product is quickly replaced by a superior technology embodied in another. Think of Nokia and Blackberry and Apple and Samsung; AOL and Google; Kodak and Cannon or Nikon. In televisions the path has been HD to 4K to 3D to HDR video. Much of this chapter is devoted to trying to answer the question: "Why is there a gap between best available and commercially adopted HDR technologies?"

A complete HDR pipeline is a system made up of key components: (1) cameras capable of capturing at least 14 f-stops of luminance range, fully visible RGB color gamut (0.005–10,000 cd/m^2); (2) standards or formats for HDR signals off the camera, for converting data files into a video shown on a display; despite 3 years of MPEG discussion, this is still a contentious element; (3) programming and content provision in the HDR format; (4) transmission or distribution along a coaxial, fiber or wireless network capable of delivering the HDR compressed signal to the household; and (5) televisions capable of receiving and displaying the HDR video image with a peak brightness of, say, around 6000 nits.

We need to define terms. Following the work of COST Action IC1005 [5] and adopted by MPEG ad hoc committee on HDR in their Call for Evidence [6]:

Standard dynamic range (SDR) is $\leq$10 f-stops (aka low dynamic rage (LDR)).

Enhanced dynamic range is $>$10 f-stops and $\leq$16 f-stops.

HDR is $>$16 f-stops.

Converting f-stops into units of dynamic range 16 f-stops equates with contrast ratio of difference of $2^{16} = 65{,}536{:}1$. This is normally noted as 100,000:1; approximately what the eye can see in a scene with no adaptation. Throughout this chapter HDR is taken to be associated with 32-bit images.

The system is disintegrated, in that no single firm takes responsibility for executing and controlling all tasks, products, or processes in the HDR system or pipeline. This has implications for each individual firm's decision as to whether or not to adopt and invest in any one of the system components. One implication is whether the costs to business of investing in commercial exploitation of best available HDR technologies exceeds what they see as any plausible future revenue stream. Another is an ex ante assessment of whether consumers will pay the kinds of price premiums for a higher 16- or 32-bit HDR television that are implied by the prices of existing SIM2's 4000 and 6000 nit displays and Dolby's 4000 and 2000 nit monitors; named Pulsar and Maui, respectively.

A third implication for decision making is one firm's decision of whether or not to invest in best available HDR technologies is conditional on future, often uncertain, expectations of other firms' intentions. This does not have to be simultaneous. Camera manufacturers supply cameras that can capture greater ranges of luminance and color gamut than the typical television or cinema projector can display. When America switched from black and white television to color, the studios and broadcasters created a library of color film and broadcast footage long before color televisions were sold in significant numbers [7].

It is conditional because of the need for technological and product compatibility; both backward-looking and across contemporary, and future technologies. Without compatibility the incentive for a firm to invest and adopt is much reduced. The risk of investing in a technology that is not backward compatible, for example, is the value of past investments trends to zero. They are left stranded by superior technologies if the functions of past technologies cannot be incorporated in new products. Producers and consumers can suffer from stranding.

Compatibility is helped by a coordinating mechanism among firms. Coordination makes it more likely that investments are made in complementary technologies. The problem is the costs and benefits of coordination may not be neutral across interested businesses. Coordination through exploitation of market power by leading producers sponsoring a standard, for example, usually results in the accumulation of significant license income (excess profits) to the sponsor to the detriment of other, license buying, firms. Coordination through the creation of standard by a committee, for example, MPEG, risks selecting a standard that becomes technologically inferior soon after it is adopted, or is not seen as commercially viable option by enough firms for it to be widely adopted. Coordination by government regulation that imposes a mandatory standard on the sector, risks rewarding many winners but penalizing many losers whose products and technologies do not conform. Losers usually complain or go to court; winners keep quiet.

These influences on a firm's investment decision form the basis for assessing the likelihood that the current commercial HDR adoption will converge with best available HDR technologies. There is, though, one more factor that is key to thoughts on adoption. The market for HDR video has the characteristics of a network. When one firm with a relevant technology or product enters the market, other firms supplying complementary technologies all benefit. When one new buyer buys a particular kind of consumer electronic device, all other consumers with compatible devices also benefit from this network expansion. All producers hope to gain from larger scale of production. Greater scale lowers unit costs of production. Reduced costs feed through to lower prices of products in the shop or online. The logic of this line of argument is adoption on a large scale and is more likely if there is one network of complementary HDR technologies in the HDR system. This is rather than many competing networks based on incompatible systems. With competing systems, producers associated with different networks hope to attract more customers, and through this, create a bandwagon of more and more customers joining their network. Ultimately a firm's aspiration is market dominance for their chosen network. Securing industry-wide agreement to a specific standard or format for HDR signals is often a first step to a particular network assuming market dominance.

The remainder of the chapter is organized as follows. Section 2 examines the technical characteristics of current HDR television adoption, asks what is new and explores reasons for this kind of adoption, rather than something more radical. Section 3 asks whether current adoption is likely to be short-lived and a step to adoption based on convergence between commercial application and exploiting best available HDR technologies. This is seen from the point of view of cost and other constraints on producers of switching to a 16-bit, and then a 32-bit pipeline. Section 4 reviews the significance of the need for coordination and compatibility within the context of competing networks. Section 5 explores how compatibility and coordination might be brought about, with particular emphasis on contrasting action through standards promoted by sponsoring firms, or agreement through a committee. Finally, Section 6 concludes by offering some observations on prospects for further adoption and whether what we currently experience is likely to endure.

By definition, we are attempting to analyze a world that does not exist. This is the widespread future adoption and dissemination of 32-bit based HDR video on a variety of consumer devices and in cinemas. We believe there is value in drawing on evidence from recent developments in the market for phones, tablets, laptops, and desktop computers and to a lesser extent, the adoption of color televisions, albeit many years ago.

2. CURRENT HDR VIDEO TELEVISION ADOPTION: WHAT AND WHY

2.1 Incremental Innovation

The adoption of recently launched HDR televisions represents an increment of innovation, mainly in the application of full array LED backlighting on a commercial scale. Of greatest significance is reducing LED heat and power consumption, so that television running costs approach levels consistent with profitable sales. Successful innovation leads to profitable sales based on the combined value of enhanced viewing experience of HDR video over SDR, and the price premium charged for the former over an equivalent sized 4K television. In other respects, for camera, distribution, and compression, maximum use is made of existing IT transmission infrastructure. (Fiber, wireless for hardware, share of spectrum; a software processing languages consistent with 10 or 12 bit pipelines.) The world's major suppliers of televisions have overcome a perennial question of early adopters. Is it better to go first, or wait and learn from others' mistakes? They unambiguously have gone first.

Arguably the greatest novelty is in the peak brightness of the new consumer HDR displays. This is around 1000 nits, or twice that of conventional SDR televisions. Contrast ratios on the larger commercial HDR television are thought to be around 2^{11}:1 compared to about 2^{8}:1 on a flat screen SDR television. Highlights are brighter. Frame rates are up and there is a lot of research to improve the color gamut, including the use of quantum dot technology [8]. The amount of enhanced viewing experience that comes from watching HDR content is scene dependent. Scenes with bright sunlight, white snow, and dark forests inhabited by bears and wolves lend themselves to HDR because of the wide range of light: bright to dark. This is also true for scenes where there is mystery lurking in dark areas, and where the unveiling is in the detail of grades of gray areas. Daylight scenes with a narrow range of light, and few areas in shade offer less benefit from HDR. Images of fields of sunflowers or bright red roses show off the vibrancy of color far more in a 4K HDR video than an LDR HD sequence. In these senses newness has very variable relevance to the viewer. As Fig. 1 shows, daylight scenes with a narrow range of light and few areas in shade offer less benefit from HDR than a candle in a dark room.

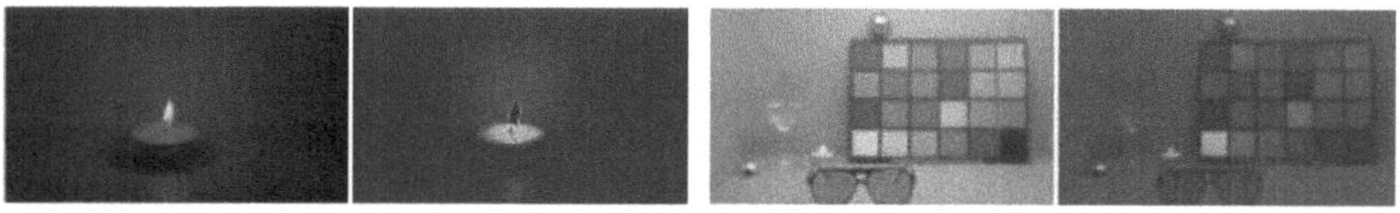

Fig. 1 A single candle lit in dark a room has a dynamic range 16.8 million to 1, while items evenly lit by sunlight only have a dynamic range of 256 to 1 [9]. *(Image courtesy of Jassim Happa.)*

When Dolby have shown off their state-of-the-art 4000 nit higher dynamic range monitor at IBC or NAB exhibitions, they have done it in a dark room. They are mimicking the light conditions of a commercial or home cinema. The market for their monitor is the studios and other movie content makers. It is not surprising that they base their marketing on replicating a viewer watching a movie in a cinema. Home cinemas typically are in rooms 4 × 8 m [source SIM2]. Dark rooms in a home might be the conventional viewing conditions of people who buy the latest Sony, Samsung, Panasonic, or LG television. This seems unlikely given the investment these businesses have put into widespread marketing of their HDR televisions in specialty shops and department stores. But what household in Europe or North America living in houses, apartments, or condominiums is going to be able to accommodate an 85-in. television in their sitting room? Perhaps HDR TVs are not aimed at a mass market of existing television owners seeking to replace their aging televisions.

Viewing in a dark room extenuates the perceived contrast on the screen and offers an enhanced viewing experience. It differentiates an HDR television from an SDR one to a point where it is easy to see consumers being prepared to pay a higher price for it. But if the room lighting is approaching day light, so that the contrast is less stark, would the price premium be anywhere near as much? A recent survey of a number of consumers of HDR TVs has shown that in order to clearly see the benefits of HDR video backlit LCD, displays should be viewed in ambient lighting of <5 nits, while OLED HDR displays with their deeper black levels should only be viewed in a dark environment, in which walls are not painted white [10].

Despite significant work for more than 3 years by the MPEG ad hoc committee on HDR, there is as yet (Jul. 2016) no single agreed standard for HDR. A number of televisions, including Sony, Samsung, and Panasonic use HDR10, while others, such as LG, use both HDR10 and Dolby Vision to deliver HDR [11]. Common to both these methods is the Perceptual Quantitiser (PQ) method of HDR video encoding [12]. PQ is derived from the Barton curves of Barten [13], so not that new. Nor are the methods used to distribute content from capture to household display new. The larger bit-depth of 10 or 12 bits, needed to facilitate the retention of more data and create the digitized HDR image is a modest increase on that required by an 8-bit SDR television. Video codecs for distribution use uniform standards such as H265 or high efficiency video codec (HEVC) [14].

While there is commercial novelty in the widespread marketing of new UHD/HDR televisions, the technology that creates the HDR video display image is older. A technology using a full array matrix of either white or colored LEDs to backlight the front television panel was presented by Brightside in 2004 [15], and further developed by Dolby and SIM2 after the acquisition of Brightside by Dolby in February 2007. What is new, compared to the original approach of Brightside, is the use of LEDs around the edge of a display that throw directed light onto mirrors at the back of the panel.

The distribution of HDR content from camera (as raw uncompressed linear data or tone map compressed) requires conversion into smaller file sizes so that distribution, and then creation of an image on the HDR display is more manageable and cost-effective. This is not new either. All the new HDR televisions receive data files in this way and convert them to create the desired image using the PQ curves [12].

4K HDR televisions available complement, rather than replace existing producers' range of 4K or full HD televisions of different screen size. They do differ in internet capability. Some are branded as SMART TV, others Android. The price difference between 4K and 4KHDR, for otherwise the same specification television, is around double. Prices of Samsung 65-in. 4K and 4K with HDR televisions are £1400 and £2700, respectively, in a well-known UK department store. The difference between the prices of 48-in. TV full HD compared with 4K HDR television was just under £1000. Prices were £470 and £1400, respectively. These are large price premiums. It is not known how many HDR 4K televisions have been sold.

2.2 Why Might Any Form of HDR Video Imaging Be Adopted?

At its most basic, the recent adoption of HDR televisions can be explained in terms of television manufacturers innovating in response to pressures caused by falling retail prices of TVs, short- and long-term. The average quality adjusted price of televisions in the United States has declined by around 90% since 1997, according to the US Bureau of Labor Statistics [16]. The price of a 32-in. LCD television fell from $725 in 2008 to just $200 in 2013 [17]. Another market research report, this time for the United Kingdom, shows the price of 4K televisions declining from £7615 in Nov. 2013 to £1314 in Jul. 2015.

2.2.1 Response to Falling Prices

Price falls have not been uniform across all types of televisions. Larger television prices have held up better than small screen televisions. However, the price of a larger television at its launch is higher than the same television being sold a year later. There is nothing unusual in these price trends. They follow a pattern of price falls in a range of consumer electronic devices, including phones and laptops. To take one recent example, the price of an iPhone 6 (64 Gb) dropped from £699 at launch to £620 12 months later. Samsung Galaxy 6 (64 Gb) prices declined even further [18].

Some of the price falls in televisions reflect structural shifts in demand away from televisions to watching video on computers, laptops, tablets, and phones. Following technological convergence across viewing formats, the numbers watching televisions has also been affected by the relative decline in the preference of households to watch films or sports via membership of a cable network.

2.2.2 Rapid Technological Change

Turning to the competition; mobile devices, the first iPhone launched in 2007 had a 3.5 in. display, a resolution of 320 × 480 (HVGA), and graphics processor of 103 MHz. The camera was 2 MP and captured at 30 fps. The latest iPhone 6 plus has a screen of 5.5 in., a resolution of 1020 × 720 and a graphics processor of 450 MHz. The camera captures 8 MP at 60 fps. Whatever the device that shows the video footage to the viewer, the effect has been the same. There has been a marked improvement in the visual image, in color quality, in resolution, in sharpness of image, smoothness of movement, the absence of ghosting, and in observed detail. Whether the video is to inform, entertain, excite, or change moods, innovation in the use of new imaging technologies makes their achievement more likely.

Combined with advances in communication technologies, access to watching video is no longer as restricted by location. Few in developed countries have to go to a cinema or stay at home to watch a film.

Distribution of content is no longer to dedicated fixed points where people can watch it. Rather than consumers moving to where the content is, the content is now sent to mobile devices where the consumers are. Increasingly it is made available when the viewer wants to watch it and not when the broadcaster or cinema wants to show it. Widespread past adoption of new imaging technologies has widened and deepened the market for, and supply of, video content. Think of this in terms of the number of electronic devices capable of displaying video content, diversity of consumer electronics having a functional ability to show off moving images, or total viewing hours on all types of products. Adoption of HDR video is part of this trend.

The conventional explanation of these rapid technological changes in a video pipeline combine observations about improved camera sensors, continued relevance of Moore's law, the ability of semiconductor designers and manufactures to squeeze ever more computing power on smaller and smaller processors, more efficient file compression, and marked reductions in power consumption used by LEDs of a given brightness. Seen in this way, technological change in video is innovation led on the back of greater integration of hardware and software.

2.2.3 Process of Adoption

Historically, technological innovation to market exploitation and success involved extensive experimentation, validation, and evaluation by selected early adopters. More wide-spread technological diffusion followed once lessons learned by early adopters had reduced technology and market risk. Risk reduction stemmed from revision of what customers would accept and product adaptation reflecting a more precise understanding of customer requirements, and the price they were prepared to pay to have them satisfied. The learn-adapt cycle was repeated until either market maturity was reached, or commercial failure was evident, and the innovation was written off.

This is a poor descriptor of adoption and diffusion of new technology in consumer electronics. Here product cycles are measured in months, rather than years. Payback periods are typically 2 years, rather than 5. Benefits from adoption of consumer electronics devices are very evident to an ever-increasing number of consumers very early on.

The past adoption of video technologies or systems occurred because technological change was in things which mattered to producers or consumers. Speed of technological change creates its own dynamic. Successful advances today, based on accumulation of new knowledge today, encourage further incremental change and innovation tomorrow. Momentum in the adoption of new video imaging technologies is neatly illustrated by the very short product life of MP4 players. They were very soon surpassed in their functionality and utility by smartphones that bundled up video downloads with audio, and a mass of other applications. The emergence of HDR televisions is one aspect of this technological dynamic.

2.3 Why Is Adoption at the Moment Taking the Form It Is?

A related question to this is: "*Why now?*" Possibly because all other viable ways of differentiating televisions have already been tried, for example, 4K and 3D. Arguably both were easier and quicker to implement than HDR displays, because the required technological advance came quicker. The time lag between technological development to go from SD (640×480 ppi) and HD (1920×1080 ppi) was 38 years. From HD to 4K was much shorter; at 11 years, while from 4K to HDR, 3 years.

Early HDR displays had technological deficiencies. They were power hungry, had very noisy fans, and needed significant amounts of piping containing liquid coolant to help reduce the heat given off by the array of numerous LEDs. No more; the latest 65-in. HDR televisions consume as much power as a 32-in. HD television 4 years ago: around 120 W.

2.4 Is It a Form of Adoption That Will Persist, or Be Quickly Replaced by Something Better?

Current HDR video commercial adoption is less than technologically optimal. Defining commercial adoption of HDR video in terms of 10 or 12 bit images is closer to terminology usually associated with SDR. HDR imaging offers the potential of capturing the full range of lighting in a scene and delivering it in a digital format along a pipeline to a display. Known as "scene referred," this full range of lighting can be kept, while maintaining physical accuracy, if 32-bit IEEE floating point values are used to represent each color channel. This means that 96 bits per pixel (bpp) are needed, compared with a standard image of just 24 bpp [19]. A single HDR frame of uncompressed 4K UHD resolution (3840×2160 pixels) requires approximately 94.92 MB of storage, and a minute of data at 30 fps needs 166 GB. This is currently prohibitive on existing ICT infrastructure. Efficient data formats and compression techniques are thus essential if HDR video is to be widely adopted [9]. This is partly semantics. Its relevance to the issue of HDR video adoption is 10 or 12 bit imaging is significantly below what is technologically possible, and a long way from the view that true HDR represents imaging that matches the human visual system.

2.4.1 *Focus on Costs of Switching and Stranding*

Why 10 or 12 bit images and not higher? One factor that helps explain the adoption of 10, or 12 bit HDR video, is it overcomes the problem of stranding. Stranding occurs where the adoption of a new technology makes all earlier technologies obsolete in technical and commercial terms. Under conditions of stranding, all previous sunk investment in all parts of the video pipeline would be worthless upon the adoption of a completely incompatible with old, new technological system. This is not an attractive proposition for firms involved in current video production of hardware, software, and content. Having sunk significant investment in current TV technologies, it is in their vested interest to adopt technology that is backward compatible. This is an HDR system that complements 4K and 3D, and uses existing IT communications infrastructure capacity. Much more important is it means the possibility of significant financial benefit for those businesses that can switch to 10 or 12 bit imaging at a modest extra cost. The avoidance of stranding introduces a bias in favor of the status quo. Consumers can also suffer from stranding. UK Broadcasters switching off the analog signal is a recent example of the risk that a consumer might be left stranded.

3. CURRENT ADOPTION: SHORT TERM OR ENDURING?

3.1 Is This Kind of Adoption Likely to Persist, or Be Overtaken by a More Impressive Technology?

Current best available HDR video technologies are far more advanced than 10 or 12 bit imaging. A 16-bit HDR pipeline is perfectly feasible. A 32-bit pipeline has been shown by academic researchers at IBC and NAB. Capture of at least 14 f-stops is possible from a range of commercially available cameras ARRI, Sony, and Grass Valley. There is no need to tone map on the camera. In the current workflow, this footage is then mastered for 1000 nits and encoded using ST.2084. For high bit rates, the file format used is ProRes 422HQ. Other HDR video codecs, both one-stream and two-stream could be used [20]. Distribution is facilitated by H265 or HEVC compression. Existing Dolby HDR monitors, such as the Pulsar and Maui, although not commercially available, have peak luminances of 4000 and 2000 nits, respectively. The latest SIM2 displays, on the other hand, have a peak luminance of 6000 nits with a 10,000 nit display due to be shown at IBC in September 2016. At NAB in 2015, goHDR, Vicomtech, ARRI, SIM2, and the University of Warwick demonstrated live streaming of HDR video from an ARRI Alexa to a SIM2 display. While this clearly demonstrated the potential of HDR to provide a step change in viewing experience, the large data requirements of HDR video means that for remote broadcast of HDR video, the quality is compromised by the need to significantly compress the video stream to handle it on existing ICT infrastructure.

3.2 Constraints: Why Not Adoption Closer to Technological Optimum Possibilities?

3.2.1 Uncertainty Over Consumer Demand and Willingness to Pay

Would extending the dynamic range significantly above what is currently offered on mass market televisions make enough of a difference to viewing experience that as sizable proportion of consumers would pay to enjoy it? One thing to note is, currently an HDR 4K television carries nearly a 100% price premium over an equivalent 4K non-HDR enabled TV. Another is evidence on consumer response to the launching of HD televisions is supportive, although there was significant variation across age groups and income levels in both willingness to adopt and pay higher prices [21]. Early adopters of HD television were more likely to be young and have high disposable incomes. They were also well-informed about the technical characteristics of the television and could relate the benefits of higher resolution and larger screens to their enhanced viewing experience [22]. Demographics and income levels also helps to explain why successive generations of smartphones, with every improving imaging attributes, took market share off previous generations of phones. Older generations adopted HD televisions at a much slower pace. Critically they valued the gain from the new technology less than they valued the loss of familiarity with old technology [23]. Much, it seems, hinged on consumers' reference points in their purchasing decision. These reference points included the attributes and characteristics of their existing television, their interest in new technology, and whether they would be unable to watch television if they kept their existing set [24].

None of these points, taken on their own, are likely to stimulate huge industry investment in much higher dynamic range televisions. There is too much risk and too little prospect of reward. And one interpretation of the arguments put forward by proponents of the 10 or 12 bit image generation, is there is no need to increase the bit depth because the viewer could not tell the perceptual difference between videos shown at higher bit depth [12]. However, not everyone agrees [25].

The risk reward tradeoff could be improved by a focus on reducing cost constraints of 16-bit HDR adoption. Two constraints are of particular importance. First is the cost of 4000 nit displays. A SIM2 47-in. display sells for around £19,000. A 4000 65-in. Dolby Vision reference monitors is thought to cost around $60,000. The second is the cost of firms switching from 10- to 16-bit pipeline.

3.2.2 Display Costs

Prices of displays could, in principle, come down if costs of components and manufacturing production came down. This might be because of (a) technological improvements in the LEDS, reduced voltage and current for the same level of brightness, or intensity of RGB colors (peaks or duration) or the means of lighting-specific pixels on the display: individual pixels or in clusters, (b) falling costs of LEDs of a given quality, (c) in the production of the (LCD) front panels, and (d) the manufacture and assembly of HDR televisions.

Samsung claims that their 1000 nit HDR television (Samsung UE65KS9000) consumes 118 W or an annual consumption of 164 KWh; 118 W equates with an average video luminance of around 30% of the peak luminance, and that the peaks are present in the video for only 5% of the total time. Applying a similar estimation method to a SIM2 4000 nit display gives rise to an annual consumption of around 350 KWh. Significantly higher certainly, but not four times higher. And much less than an estimated 463 KWh that a 42-in. plasma screen consumed in 2009 [26]. The innovative power management and heat dissipation technologies present in the SIM2 displays means that for any given level of luminance, SIM2 technologies, in fact, use less power and generate less current and hence, heat than those technologies embodied in the Samsung television.

SIM2 monitor manufacturing is labor intensive and carried out in a high-cost country. Conventional SDR televisions are also manufactured using labor intensive methods, but they are made in low-cost countries, in very large volumes, and at ever-decreasing unit costs of production. This suggests costs of HDR displays could come down by shifting production and assembly to low-cost countries and producing on large scale. This has been the experience of production systems used in the manufacture of a range of consumer electronic devices. Costs of producing Apple smartphones, for example, have fallen because of disintegration of supply and value chains. Costs fell because of declining transaction costs at different technology interfaces and the separation of production processes and supply of key components. Unbundling processes into separate tasks allowed sourcing from the most competitive supplier. Most importantly of all, costs were reduced by the production of components in very large volumes. This meant reaping a major benefit of economies of scale. Unit costs were further reduced because of firms learning to do manufacturing processes more efficiently.

3.2.3 Switching Costs

Firms that supply the technologies that facilitate the creation and showing of video incur switching costs if they move from a 10- or 12-bit to a 16-bit pipeline. Significant investment is required to switch from, say a 1000 nit screen, to one that gives out four times as much peak brightness. A 4000 nit display, however, only has value if other transmission technologies facilitate the transfer of data into pixels and then a video. Costs of switching to a transmission system to accommodate 16-bit files are significant. In part, this is because where bandwidth was at full capacity, 16-bit allocation of time banded slots would mean, either fewer slots, longer time slots, or slower transmission speeds. With scarce bandwidth, this would mean higher prices charged by those that own the distribution channels to facilitate transmission of bigger 16-bit files. There would also need to be significant investment in design and manufacture of decoding semiconductors and software that runs on them. Then there is a chicken-and-egg problem. Investment in 16-bit transmission infrastructure is more likely if there is sufficient content to justify it. Investment in content is more likely if there is confidence transmission to the home is in place.

It seems highly unlikely that the costs of switching to implement a 16-bit HDR system in either absolute terms or relative to current costs would be the same for producers at each stage of the video pipeline. The switching costs for camera producers would be low because 16 f-stop cameras can already capture most of the color gamut and luminance range associated with 16-bit HDR video. If firms in some parts of the pipeline do not switch, adoption is unlikely. This is because adoption is beneficial only if all in the HDR video pipeline adopt. If firms in one part of the HDR pipeline fail to adopt, no adoption of the whole system can take place. The default inert option is all interested firms do nothing until there is greater certainty over future costs and new technology application.

Table 1 Comparison of smartphones

	iPhone 6	Galaxy 6
Display size	4.7 in.	5.1 in.
Resolution	750 × 1334 pixels; 326 ppi	1440 × 2560 pixels; 577 ppi
Camera	1080@60 fps or 720@240 fps	2160@30 fps; 1080@60 fps; 720@120 fps
Battery	Li-po 1810 mAh, 6.9 wh	Li-Ion 2550 mAh

3.2.4 Important Qualification

In time HDR video could appear on other devices/platforms if there was effective technology transfer from HDR televisions. This raises the obvious question as to how consumers might respond to this. There is some anecdotal evidence that LED brightness and reduced power consumption in smartphones might be expected to carry greater weight than improvement in sensors to enhance image capture, for example. Reading forums that compare Samsung and Apple smartphones and tablet technologies, it is clear that the technologies that go into each device are not equal in technological efficiency, see Table 1. However, in terms of what is the key technological differentiator in the eyes of the buyer/final customer, bloggers, or online reviewers' ranking of technologies according to relevance to customer experience puts display/screen ahead of the processor, battery, and camera?

But there is a puzzle. The Galaxy has a far superior image display set of technologies. If quality of video on the screen is important to the customer, it might be expected that Samsung would charge a higher price for the Galaxy 6. Unencumbered by contracts, their respective retail prices were almost the same, at around $750 for a 64 Gb phone. Perhaps there are compensating differences on the iPhone, for example, design or number of apps [27].

4. SUPPLY CONDITIONS AND HDR ADOPTION

4.1 Relevance of Networks

The markets for televisions, smartphones, tablets, DVD players, and a range of other video devices have the characteristics of being networks of consumers and producers. The conventional examples of the benefits of consumer networks are postal networks and more recently, social networks. In each case, when a new person joins the network, everybody in the network gains from having access to an additional subscriber or user. Those using Gmail gain from millions of others using the same email system. A consumer may be reluctant to join Gmail if everybody else used Microsoft Outlook.

Under conditions of network benefits, one supplier of an element of imaging technology gains from other suppliers offering complementary products. Content producers and broadcasters using the same format gain from knowing this means distribution to a larger number of customers. Consumers gain every time another consumer buys a television, tablet, or phone that has the same operating system as theirs. As the market for content expands, so the costs of supplying one unit of content falls, because of economies of scale or scope and thus, creates the possibility of lower prices and higher sales. Indeed consumers may join a network because they believe that by doing so, product prices will fall in the future. Fragmentation and incompatibly between competing systems and formats lessens these network benefits. The point about networks is, once created, they can bring about bandwagon effects so that those joining the network increases the benefits to all. Under circumstances of common interest to all, the greatest benefits come from a single integrated network available to all [28].

4.2 Compatibility

What is important about networks is that the products that people buy to exploit the benefits of being part of a network are compatible. Few technologies are valuable in isolation. Their value comes from the way they are combined with other complementary technologies [28]. In HDR the obvious point is content needs to be captured in ways compatible with distribution and delivery across a range of different displays. Software in one part of the pipeline needs to be compatible with software at the interface at another part of the pipeline. Software needs to be compatible with the hardware it runs on. All producers in the capture to display pipeline know compatibility is essential; incompatibility is road to isolation and ruin (Fig. 2).

Compatibility is fundamental to explaining how firms in consumer electronics found all kinds of ways of getting hardware unit costs down. *Why and how?* Products of the consumer electronics sector were characterized as bundles of technologies that were compatible. This meant that individual products had little value on their own. Their value rose when they combined with other products [28]. The value of each individual product reflected the most advanced technological capability of the business that

Fig. 2 There is money to made if all the components of the HDR pipeline are in place.

produced it. This, in turn, stemmed from firm investment in technology-specific research and development and an in-firm tacit understanding as to how best to turn knowledge of new technology into successful products. Costs were reduced by exploiting differences in firm specialization in technologies associated with capture, signal distribution, and image display. Multiple firms produced a multiple variety of products.

Cost reductions in the supply of key consumer electronics components were assisted by modularization in component design and manufacture. Modularization meant that components could be sourced from the most efficient supplier. However, the benefits of modularization largely arose because there was compatibility between components, functions, and software applications. Compatibility increased the scope for a single component to fit in a single product or in numerous products. Greater compatibility in breadth and depth among components and software in laptops, phones, computers, and tablets increased the possibility for higher volumes of individual components and through this, much reduced unit costs of production. It is not obvious why such an approach based on falling unit costs with ever larger volumes could not be applied to the HDR system.

The HDR pipeline from capture to display is disintegrated. No single firm owns or controls all parts of it. For individual firms engaged in different parts of the pipeline, the conditioning of their individual response to the prospect of say 16-bit HDR is likely to hinge on their views about areas of major divergence in compatibility of today's video technologies compared with tomorrow's. The important consequence of this is that one firm's possible future benefit from adopting higher HDR is conditional on others acting in complementary ways. Under these conditions of future uncertainty of supply, no firm gains from acting alone. But they can also lose from incompatibility. One example is a firm was locked into producing a product with an outdated technology. There are lots of past examples. Analog versus Digital TV. Mini discs against music downloads; further back, AM versus FM radio. Some consumers will therefore, be left with equipment that is incompatible with the winning system.

It is easy to make the general case for the need for compatibility. It is much harder to say how it might be applied in practice at some future date. Diversity of product markets for video, and different types of content directed at different audiences complicates compatibility. The significance of variance in scene contrasts and choice of scene selection opens up a divergence of interests between broadcasters and studios among major content suppliers. Many of the examples of where true, scene-referred, HDR would have significantly improved viewer experience come from live sports broadcasts; anywhere where there is bright sunlight and dark shade in a single scene. For example, one part of the football pitch is in sunlight, the other in shade caused by the shadow of big stands. Or the players on the pitch are in shadow, while the fans on the terraces are in sunlight. A goal is scored and the fans' reactions are shown as a mass of white pixels. It is possible different interests in content will produce the same firm incentives to invest in an HDR pipeline, but it seems unlikely.

At one level, studios and broadcasters can anticipate each other's general interests in HDR. At a more detailed level, firms producing complementary elements of the HDR video pipeline for both broadcasters and studios, would struggle to say with any certainty the relative proportions of these different content producers' interests. As important to an investment decision is estimating the magnitude of volume to be produced of these complementary products, and their timing.

To complicate matters further, firms must calculate how much of the value generated by the sales of HDR video systems might accrue to them. Would it matter if the content came from Netflix or Amazon, rather than a conventional broadcaster or Hollywood studio? At the time of writing, the value of streaming or downloading SDR video to tablets and phones accrues mostly to the intellectual property owner of the brand. Apple, for example, received around 60% of the total value added generated by iPhone 4 [29].

Differences in compatibility could come at technology interfaces, in conflicting scope of intellectual property protection, and divergence of view as to how a combined technology might be implemented. The adoption of one sponsored technology might result in less redundancy of a firm's own technology than another competing technology. The value, in application, of the combination of one's own technology and a sponsored technology could vary across sponsored technologies; if for example, the attraction of some products increased consumers' willingness to pay higher prices, but to a varying extent.

If HDR video were to be applied to a variety of display devices such as phones, tablets, laptops, and televisions, each supplied by different producers, a system which is best suited to one device and one set of customer preferences may be less than the most profitable if applied to another device. Differences in power requirements on a phone compared to a television could force producers to adopt different HDR video systems, for example. Instead of uniformity HDR system, we would have variety.

Any sensible investment appraisal made by a firm in the HDR pipeline ought to try and reconcile and give weight to these different elements of future uncertainty on the form, direction, and timing of product compatibility. And they need to confront a chicken-and-egg problem. Investment by one firm in a particular technology is only worthwhile if it knows other firms contributing to the same technology system are also going to invest in creating complementary technologies.

5. COORDINATION AND THE ROLE OF TECHNOLOGY STANDARDS

5.1 Role of Standards

Compatibility requires coordination among interested producers. In consumer electronics, generally coordination has taken the form of industry agreement to common interface standards. A standard can be defined as technical specifications that allows one technology to be compatible with another at a specified interface. Their main purpose is to coordinate diverse actions of numerous firms that produce products that complement each other. Standards exist because groups of firms acting together can achieve greater individual and aggregate benefit than they can by acting alone. By agreeing to a standard, firms aim to reduce uncertainty over the path of a particular technology and minimize costs of developing technological redundant (i.e., incompatible technologies) through the generation of competing standards.

5.1.1 Factors Promoting Adoption of a Common Standard

There are a number of important factors:

- One standard's superiority in functional/performance terms compared to others.
- Very small cost to users of incorporating technology implicit in the standard on existing hardware of current and earlier vintages.
- Very small cost to users of integrating technology implicit in a standard with other computer code or language, that is, a high degree of interface compatibility.
- Very small cost to users of switching from current technology or prevailing standard to new superior standard, that is, how far is a new standard generic suitable to widespread application on a variety of hardware and combined with a variety of other programming languages?
- Significant incremental value in the product produced that embodies the new standard; to producer and user (e.g., quality of video image, attraction of HDR in image).
- Flexibility or adaptability to novel future technologies, such as hardware and applications.
- Significant network benefits.

5.1.2 Factors Impeding Adoption

These include:

- High costs of bringing about compatibility between technologies embodied in a standard and current hardware and software.
- Dominant firms in HDR video pipeline seeking to protect their own proprietary IPR; they create barriers to new standard adoption.
- Absence of backward compatibility.
- Limited take up in absence of clear benefit in application.

5.1.3 Options to Creating a Standard

The standard that is causing the greatest contention is an HDR video standard for the file format or video compression and decompression, when data files enter (from the camera) or leave (to the display) fiber or wireless transmission.

Option 1 involves efforts to find a way of interested industry agreement to a single HDR standard and have embraced most of the possible options to creating a standard. The MPEG option reflects an approach based on collective collaboration through a voluntary standards setting organizations (SSOs) where all interested parties seek consensus on an agreed format or interface technology. They have thus far failed.

Option 2 is large firms that are technology leaders act as standard setters working as a consortium. HDR 10 is an example of such an approach. It is an open standard. At the time of writing, Jul. 2016, it is being more widely adopted than Dolby Vision; a competing closed standard. Both formats are complemented by a separate standard agreed among broadcasters.

Option 3 is sponsored promotion of a particular standard by a single firm in the expectation of significant license income. For license fee payers balancing their benefits from having a single standard with the costs of being locked in to a particular supplier of Intellectual Property is challenging. Dolby Vision is an example of a firm-sponsored standard where take up is controlled by Dolby in the expectation of license income.

Option 4 is market competition or standards wars where different technologies and the firms that seek to promote them hope that their technology will be attractive enough to encourage firms to create a bandwagon. This approach, however, can be high on intensity, costly in duplication, and maybe inefficient, if numerous incompatible standards coexist; inefficient to a point were significant network benefits are forgone. Technological leaders do not have to be dominant firms, but they need something firm-specific that makes their technology superior to others [30]. Thus far, option 4 has not featured in ways of generating an HDR standard, although there are two HDR standards now available: SMPTE ST2084 [31] and ARIB STD-B67 [32]. Central to both of these methods are perceptually based transfer functions. In SMPTE ST2084 this is the PQ curve [12], while ARIB STD-B67 uses Hybrid Log Gamma (HLG) [33].

5.2 Some Observations on MPEG Progress

What might be the explanation of the lack of MPEG progress? Past experience of standard setting by committee suggests failure thus far to find agreement is a function of:

(a) The number of firm-sponsored technologies submitted to it. A firm's preference will be largely conditioned on what it considers to be the costs of switching from an existing technological path to a new one, and the extent to which it holds the intellectual property rights, rather than having to pay for a license to somebody else. So it (the firm) is likely to sponsor a proprietary standard and resist somebody else's proposal. The tradeoff is the extra profits (economic rents) each interested firm expects to make if its preferred standard is adopted, and the less than ideal amount of profit or loss it expects to make if an alternative technology is adopted.

(b) The extent to which it is possible for sponsoring firms to form a coalition by aggregating all proposals into a single closed standard. This might balance some vested interests, but create a hybrid standard that is inefficient as a means of coordinating firm actions.

(c) The extent to which it is possible for the sponsor of one technology to secure consensus from others by a system of side payments. However, paying off other vested interests must, in turn, be paid for; perhaps by a higher license fee.

(d) The number of unsponsored technologies lodged with the MPEG, often by academics. Here standard selection is a reflection of the decisions taken by enough firms in a market to create a bandwagon.

(e) The technical superiority of one technology over others according to agreed industry objective and subjective measures of performance, and whether it impedes or accelerates adoption because it overcomes or reinforces inertia between existing technology and a proposed new one. A related issue is how firms are to deal with their own or other's stranded assets. A number of comparisons of different methods are appearing, for example [20]. One clear challenge is the current lack of robust objective metrics for HDR video, and thus time-consuming subjective metrics are often required to clearly understand the differences in approaches [34].

(f) The speed of technological change which could either delay or accelerate adoption, depending on whether it was felt, by firms, to be better to be locked into today's technology or wait for a better one tomorrow. All are likely to ask: "Is it better to agree now or wait and see how technology advances proceed?" While delay may carry a benefit to some firms, it carries a cost to others. The issue is whether delay ultimately makes all worse off.

(g) The size and distribution of costs and benefits across interested firms associated with the adoption of each sponsored technology.

(h) Extent of technological differences across competing submissions in both design and application. Large technological difference across standards introduces a possible cost bias in favor of some producers, but against others. Firms would be expected to minimize their own cost of switching to a new technology whist expecting others to act. Problems seem to be greatest were technological change is very rapid and costs of being locked in are significant. At a margin, a single standard is seen as more worthwhile than standards wars among competing standards. Observations based on the work of [35].

To these authors at least, it is not obvious how there will be convergence on a single standard. Probably the emergence of two standards (ST2084, ARIB STD-B67) and Dolby Vision for coding formats used in current HDR television is a sign of things to come.

6. CONCLUSIONS

The term HDR is being widely used to try and persuade customers to buy a new television. While it is clear that HDR content on one of these consumer HDR televisions *in a dark environment* is impressive, the enhanced viewing experience is less obvious in a typical daylight setting. Furthermore, the drive to provide consumer HDR televisions all require 10 bit infrastructures, which precludes being able to view the same content on the large numbers of 8 bit devices, including legacy displays, and most importantly, mobile devices. More and more video content is now being viewed on mobile devices [36]. In addition to the current limitation of 8 bit infrastructure, HDR video on mobile devices introduces more challenges, including dynamically varying ambient lighting conditions, reflections on the screen, etc. [37].

Standards for HDR video encoding are emerging, but there is already evidence that there are more efficient ways of encoding than those of the standards [9].

A theme running through this chapter is future adoption of HDR video will be a consequence of specific forms of interaction between a small number of large businesses in capture, content creation, transmission, and distribution to visualization markets. Interaction is, in part, conditioned by individual firm preference for extending the commercial value of their own past investments in imaging technologies. Past investment reflects what are seen as the basis for offering buyers particular knowledge or expertise that is proprietary, not easily copied, and is the source of significant financial benefit. To many, the ideal solution is adoption turns the adopting firm into a monopolist. In the process, the competition is either avoided or their products are rendered commercially redundant. These are supposed as being important motivations of the market incumbent firms in the HDR pipeline.

Their protection against threats from new market entrants, offering more advanced HDR video, comes from the way different incumbent firms combine complementary technologies. They bundle them (the technologies) together so that no single firm has the expertise, breadth, and depth of knowledge to offer a superior combination of new HDR technologies. Industry incumbents are also helped by very uncertain consumer response to future video content and display in 16-bit imaging. All of this creates a bias in adoption the favors the status quo.

The status quo could be destroyed. One way is because of internal conflict between firms in one or more HDR video pipelines. Within formal or informal agreements are inherent features that can contain the seeds of destruction. Perhaps expectations about the supposed future benefits stemming from the actions of others in the HDR pipeline partner firm's turn out to be wrong. Disappointment sets in, a grievance is created. Perhaps expectations were based on what turned out to be selective disclosure of knowledge offered by partner firms. Not every business need be completely forthcoming about their future intentions with respect to investments in R&D, possible future alliances, or market segments to be targeted. Keeping quiet can be the source of commercial benefit in a bilateral relationship with other businesses.

In other sectors of the consumer electronics industry, major disruption to the state of current technology adoption has come from outsiders and new entrants. Apple and Samsung smartphone technologies dislodging Nokia and Blackberry keyboard technologies is an obvious example. Could the trick be repeated with 16-bit HDR video on phones and tablets based on user capture and display?

REFERENCES

[1] ITU, Recommendation ITU-R BT.500-13: Methodology for the subjective assessment of the quality of television pictures, Tech. rep., International Telecommunication Union 2015.

[2] http://4k.com/news/netflix-has-quietly-rolled-out-4k-hdr-content-starting-with-marco-polo-season-1-13823/.

[3] http://4k.com/news/a-complete-list-of-600-titles-of-new-content-and-27-new-hdr-4k-titles-being-released-by-amazon-prime-12685/.

[4] B. Karr, A.G. Chalmers, K. Debattista, High dynamic range digital imaging of spacecraft, in: F. Dufaux, P. Le Callet, R. Mantiuk, and M. Mrak (Eds.), High Dynamic Range – From Acquisition to Display and Applications, Elsevier, 2016.

[5] http://www.cost.eu/COST_Actions/ict/IC1005.

[6] A. Luthra, E. François, W. Husak (Eds.), Requirements and Use Cases for HDR and WCG Content Distribution, Tech. Rep. ISO/IEC JTC 1/SC 29/WG 11 (MPEG) Doc. N15084, 2015.

[7] J. Farrell, C. Shapiro, R.R. Nelson, R.G. Noll, Standard setting in high-definition television, in: Brookings Papers on Economic Activity. Microeconomics, vol. 1992, 1992, pp. 1–93.
[8] Z. Luo, Y. Chen, S.-T. Wu, Wide color gamut LCD with a quantum dot backlight, Opt. Express 21 (2013).
[9] A.G. Chalmers, B. Karr, R. Suma, C. Moir, K. Debattista, Fifty Shades of HDR, IEEE Digital Media Industry & Academic.
[10] T. Schulte, J. Barsotti, HDR Demystified: Emerging UHDTV Systems Technical Paper, SpecraCal Inc, 2016.
[11] http://hometheaterreview.com/dolby-vision-vs-hdr10-what-you-need-to-know/ or http://www.cnet.com/uk/news/dolby-vision-vs-hdr10-how-to-survive-the-hdr-tv-format-war/.
[12] S. Miller, M. Nezamabadi, S. Daly, Perceptual signal coding for more efficient usage of bit codes, in: SMPTE Conference, 2012.
[13] P.G.J. Barten, Contrast Sensitivity of the Human Eye and Its Effects on Image Quality, SPIE Optical Engineering Press, Bellingham, WA, 1999.
[14] G. Sullivan, J.-R. Ohm, W.-J. Han, T. Wiegand, Overview of the high efficiency video coding (HEVC) standard, IEEE Trans. Circuits Syst. Video Technol. 22 (12) (2012) 1649–1668.
[15] H. Seetzen, W. Heidrich, W. Stuerzlinger, G. Ward, L. Whitehead, M. Trentacoste, A. Ghosh, A. Vorozcovs. High dynamic range display systems. ACM Trans. Graph. 23 (2004) 3.
[16] http://uk.businessinsider.com/historical-price-trends-for-tech-products-2015-10?r=US&IR=T.
[17] http://money.cnn.com/2011/11/22/technology/lcd_tv_prices.
[18] http://www.t-mobile.com/cellphones.
[19] F. Banterle, A. Artusi, K. Debattista, A. Chalmers, Advanced High Dynamic Range Imaging, AKPeters, 2011.
[20] R. Mukherjee, K. Debattista, T. Bashford-Rogers, P. Vangorp, R. Mantiuk, M. Bessa, B. Waterfield, A. Chalmers, Objective and subjective evaluation of high dynamic range video compression, Signal Process. Image Commun. 47 (2016) 426–437.
[21] B. Bayus, High definition television: assessing demand forecasts for next generation consumer durable, Manag. Sci. 39 (11) (1993) 1319–1333.
[22] M. Dupagne, Exploring the characteristics of potential high definition television adopters, J. Media Econ. 12 (1) (1999) 35–50.
[23] F. Sultan, Consumer preferences for forthcoming innovations: the case of high definition television, J. Consum. Mark. (1999), emeraldinsight.com, http://www.emeraldinsight.com/doi/abs/10.1108/07363769910250723.
[24] V.P. Rindova, A.P. Petkova, When is a new thing a good thing? Technological change, product form design, and perceptions of value for product innovations, Organ. Sci. 18 (2) (2007) 217–232.
[25] A. Chalmers, J. McNamee, J. Hatchett, R. Mukherjee, I. Olaizola, K. Debattista. 12 bits is simply not enough for HDR video!, BEC 2015, NAB, 2015.
[26] http://www.cnet.com/news/power-consumption-how-much-are-your-gadgets-costing-you/.
[27] http://www.gsmarena.com/compare.php3?idPhone1=6849&idPhone2=6378.
[28] M.L. Katz, C. Shapiro, Systems competition and network effects, J. Econ. Perspect. 8 (2) (1994) 93–115.
[29] K.L. Kraemer, G. Linden, J. Dedrick, Capturing Value in Global Networks: Apple's iPad and iPhone, University of California, Irvine, California, 2011, http://pcic.merage.uci.edu/papers/2011/value_ipad_iphone.pdf.

[30] J.L. Funk, D.T. Methe, Market- and committee-based mechanisms in the creation and diffusion of global industry standards: the case of mobile communication, Res. Policy 30 (4) (2001) 589–610.

[31] SMPTE, High Dynamic Range Electro-Optical Transfer Function of Mastering Reference Displays, Tech. Rep. ST 2084:2014, Society of Motion Picture and Television Engineers, August 2014.

[32] Association of Radio Industries and Businesses (ARIB), STD-B67: essential parameter values for the extended image dynamic range television (EIDRTV) system for programme production, July 2015.

[33] T. Borer, Non-linear opto-electrical transfer functions for high dynamic range television, BBC White Paper, 2014, http://hometheaterreview.com/dolby-vision-vs-hdr10-what-you-need-to-know/ or http://www.cnet.com/uk/news/dolby-vision-vs-hdr10-how-to-survive-the-hdr-tv-format-war/.

[34] M. Narwaria, M.P. Da Silva, P. Le Callet, Study of high dynamic range video quality assessment, in: Proc. SPIE, vol. 9599, 2015.

[35] P.A. David, S. Greenstein, The economics of compatibility standards: an introduction to recent research, Econ. Innov. New Technol. 1 (1–2) (1990), Taylor & Francis, http://www.stern.nyu.edu/networks/DavidGreenstein_The_Economics_of_Compatibility_Standards.pdf.

[36] International Telecommunication Union, Ooyala: Global Video Index Q3 2014, Report, Silicon Valley, California, 2014.

[37] M. Melo, M. Bessa, L. Barbosa, K. Debattista, A.G. Chalmers, Screen reflections impact on HDR video tone-mapping for mobile devices: an evaluation study, EURASIP J. Image Video Process. 44 (2015).

FURTHER READING

S.M. Besen, J. Farrell, Choosing how to compete: strategies and tactics in standardization, J. Econ. Perspect. 8 (2) (1994) 117–131.

N. Economides, The economics of networks, Int. J. Ind. Organ. 14 (6) (1996) 673–699.

J. Farrell, G. Saloner, Installed base and compatibility: innovation, product preannouncements, and predation, Am. Econ. Rev. 76 (5) (1986) 940–955.

B.H. Hall, K. Beethika, Adoption of new technology, NBER Working Paper 9730 (2003), http://www.nber.org/papers/w9730.

M.L. Katz, C. Shapiro, Product introduction with network externalities, J. Ind. Econ. 40 (1) 55–84.

M.L. Katz, C. Shapiro, Network externalities, competition, and compatibility, Am. Econ. Rev. 75 (3) (1985) 424–440.

T. Simcoe, Committees and the creation of technical standards, Unpublished working paper, University of California at Berkeley, Haas School of Business DRAFT, November 2003, citeseer, http://citeseerx.ist.psu.edu/viewdoc/download?doi=10.1.1.197.2091&rep=rep1&type=pdf.

V. Stango, The economics of standards wars Review of network economics, degruyter.com, 2004, http://citeseerx.ist.psu.edu/viewdoc/download?doi=10.1.1.200.2784&rep=rep1&type=pdf.

V. Venkatesh, Y. James, L. Thong, Consumer acceptance and use of information technology: extending the unified theory of acceptance and use of technology, MIS Q. 36 (1) (2012) 157–178.

http://www.jstor.org/stable/1816461?seq=1#page_scan_tab_contents.

J. Farrell, G. Saloner, Standardization, compatibility, and innovation, RAND J. Econ. 16 (1) (1985) 70–83.

INDEX

Note: Page numbers followed by *f* indicate figures and *t* indicate tables.

I

J

K

L

M

N

O

P

FSC
www.fsc.org
MIX
Papier | Fördert gute Waldnutzung
FSC® C083411

Kolibri 360 GmbH
Ferdinand-Jühlke-Straße 7
99095 Erfurt, Deutschland
produktsicherheit@kolibri360.de